AS EASY AS PI

AS EASY AS PI

STUFF ABOUT NUMBERS THAT ISN'T (JUST) MATHS

JAMIE BUCHAN

This paperback edition first published in 2015

First published in Great Britain in 2009 by
Michael O'Mara Books Limited
9 Lion Yard
Tremadoc Road
London SW4 7NQ

A CIP catalogue record for this book is available from the British Library.

Papers used by Michael O'Mara Books Limited are natural, recyclable products made from wood grown in sustainable forests. The manufacturing processes conform to the environmental regulations of the country of origin.

ISBN: 978-1-78243-433-7 in paperback print format
ISBN: 978-1-84317-636-7 in EPub format
ISBN: 978-1-84317-637-4 in Mobipocket format

1 2 3 4 5 6 7 8 9 10

www.mombooks.com

Designed and typeset by www.glensaville.com

Printed and bound by CPI Group (UK) Ltd, Croydon, CR0 4YY

CONTENTS

Introduction **9**

Numbers in Language **11**

Do a Number 12
Three Sheets to the Wind 12
The Third Degree 13
The Fourth Estate 13
Fourth Wall 14
Five by Five 14
Fifth Columnist 15
Take the Fifth 15
Deep Six 16
At Sixes and Sevens 17
Seventh Heaven 18
One Over the Eight 18
Cloud Nine 19
The Whole Nine Yards 19
Dressed to the Nines 20
Police Radio Codes 20
Dixie 24
Nineteen to the Dozen 25
23, Skidoo! 25
Forty-niners 27
77 28

86 28
187 29
411 29
420 30

Numbers in Fiction **31**

00000 32
π: Faith in Chaos 32
The Prisoner 33
Se7en 33
The Seven Samurai 34
007 34
8 Mile 35
8½ 35
Up to 11 36
12 Monkeys 36
21 Grams 37
Catch-22 38
The 25th Hour 39
The Thirty-Nine Steps 39
42 40
The 47 Ronin 40
Hawaii Five-O 41
Room 101 41
Les 400 Coups 42
Fahrenheit 451 42
24601 42

Numbers in Culture 43

1.618 – The Golden Number 44
078-05-1120 48
The Indiana Pi Bill 49
Numbers Games 51
Why Buses Come in Threes 54
555 56
Numbers Stations 58
The 23 Enigma 60
419 Scams 62
The Pirahã Tribe 64
The 10 Per Cent Myth 66

Numbers in Religion and Mythology 69

Seven 70
The Number of the Beast? 76
Gematria and the Bible Code 79
Modern Numerology 82
Chinese Lucky (and Unlucky) Numbers 88
Zodiacs 93
Twelve 98
Thirteen 100
Three 103
Four 105
Five 110

Numbers in Maths and Science **115**

A Mathematical Glossary 116
Divisibility Tricks 126
π (3.14159265358979323 ...) 129
The Evolution of Zero 130
Roman Numerals 137
The Fibonacci Sequence 140
i (√-1) The Imaginary Unit 142
Misleading Statistics 145
The Binary System 150
Odds and Oddities 153
John Nash and Game Theory 158
Billions 161
The Wheat and Chessboard Problem 163
Thinking Fourth-dimensionally 164
∞ (Infinity) 168

Further Reading **173**

INTRODUCTION

As you read this, the world may still be in the grip of severe economic crisis. Chances are you were woken up this morning by an alarm so as to get to work at or before a specific time. Your mobile phone is assigned its own specific number and, like all electronic devices, its functioning is based on manipulating numbers. The timing and placement of traffic lights on your way to work would have been mathematically designed to maximize efficiency, despite appearances to the contrary. Every man-made object will have been designed with numbers in mind, and many (including this one) are given unique numeric identifiers.

As Pythagoras said, numbers rule the universe. From the bizarre complexity of economics and statistics to the words and expressions we use every day, numbers are an inescapable influence on the world – even for the least mathematically inclined. It's with this in mind that I've written this book – as a wide-ranging look at the pervasive influence of numbers which, where it strays into complexity, is intended to remain as accessible as possible. For many people, maths remains a worrying and inaccessible field of study, and although this isn't a maths textbook, I have tried to make any mathematical bits fairly straightforward. A short list of mathematical terms is included (see p.116), which I'm hoping will make the maths rather simpler (and perhaps refresh memories of learning it at school).

Turning to the more mathematically inclined, I would suggest some of the books listed at the back to those looking to explore the subject in greater depth. I should also make an important disclaimer: I am not a mathematician, nor any other kind of relevant professional. I've researched the subjects within as an interested amateur, and done everything possible to make it accurate and, I hope, interesting and amusing.

In this I've had help from several sources, and I'd like to thank all the 'legendary chancers' at Michael O'Mara Books, especially my excellent editor, Louise Dixon, and Jo Wickham in Publicity. Many thanks also to Andrew Pinder and my sister, Claire, for their wonderful illustrations, the tireless Glen Saville for his typesetting work and Oli Dacombe for his vital help in checking the maths. Finally, my thanks to friends and family for their help and support as I worked on this between essays.

Ultimately, a book like this one can only ever scratch the surface of how numbers, whether culturally, linguistically or scientifically, will always affect our lives. The world of numbers stretches on before us, forever, growing as science and mathematics continue to make new strides. So here we go, from zero to infinity via Amazonian tribes, drug culture and nuclear paranoia.

NUMBERS IN LANGUAGE

DO A NUMBER

To seriously damage something or someone. The origins of this phrase are very murky, but it seems to be derived from the world of boxing, where a fighter might be instructed to 'do a number' on his opponent's face, i.e. hit it a number of times – hard.

THREE SHEETS TO THE WIND

A nautical expression, meaning 'extremely drunk'. Surprisingly, the origin of the term seems not to be sails (which, after all, are sheet-like and ought as a matter of course to be 'to the wind'), but ropes, which have a number of names depending on their function. 'Sheet' ropes control the horizontal movement of sails, and thus having three of them loose and flapping about in the wind would be a serious problem. Sailors appear to have used a system of ratings in this way, going from one sheet to the wind (slightly tipsy) to four (unconscious).

THE THIRD DEGREE

Originally, this referred to American police interrogations, which used intensive methods, sometimes including physical violence, to get an answer or confession from the suspect. Now the phrase is often used to describe any needlessly intensive or intrusive behaviour, where someone might complain of being given 'the third degree' about a past misdeed. It may derive from the membership rituals of Freemasonry, where members are graded by degrees. Admission to the third degree and the rank of Master Mason required the member to submit to an exacting interrogation ceremony.

THE FOURTH ESTATE

Derived from the 1789 Estates-General, an assembly of French citizens from three social ranks. The First Estate were clergy, the Second the nobles, and the Third the wealthiest bourgeoisie. In this highly unrepresentative system, a fourth group – newspapers and reporters – were extremely influential on ordinary French people. In Britain, philosopher and Member of Parliament Edmund Burke pointed to the House of Commons press gallery and remarked: 'Yonder sits the Fourth Estate, and they are more important than them all.'

FOURTH WALL

Originally defined by the writer Denis Diderot as the 'wall' separating a theatrical performance from the audience (with the set forming the other three walls and the stage being the floor), the term is now often applied to films and TV, particularly where a character 'breaks' the fourth wall and addresses the audience directly. The opening scenes of *Ferris Bueller's Day Off* are a particularly good example, if not precisely what Diderot had in mind.

FIVE BY FIVE

Originally a term from NATO radio-speak (see also Police Radio Codes, p.20), part of a system of rating radio signals. Signal strength and clarity are both rated on a scale of one to five, so five by five refers to the strongest, clearest signal possible. As such, the phrase (and sometimes its contracted form 'five by') is frequently used outside its original setting, usually to mean something has been understood (similar to 'crystal clear') or just to report that something is proceeding as planned.

FIFTH COLUMNIST

This phrase was, apparently, coined in 1936 during the Spanish Civil War when, in a radio address, Nationalist General Mola proclaimed his intention to take over Madrid with only four columns of troops, plus a 'fifth column' of Nationalist sympathizers inside the city who would sabotage the Republican defence. Mola's plan was unsuccessful, but the term gained popularity during the Second World War to describe possible enemy sympathizers in the Allied ranks, such as Japanese-Americans and German expatriates.

TAKE THE FIFTH

In an everyday setting, the term has increasingly been used to refer to refusing to answer a question where the truth would embarrass you. The Fifth Amendment to the American Constitution guarantees, among other things, the right to remain silent when arrested and during trial, making it illegal to force someone to incriminate themselves:

> No person shall be held to answer for a capital, or otherwise infamous crime, unless on a presentment or indictment of a grand jury, except in cases arising in the land or naval forces, or in the militia, when in actual service in time of war or public danger; nor shall any person be subject for the same offense to be twice put in jeopardy of life or limb; nor shall be compelled in any criminal case to be a witness against himself, nor be deprived of life, liberty, or property, without due process of law; nor shall private property be taken for public use, without just compensation.

DEEP SIX

Whereas one might '86' (see p.28) an unneeded internal memo, one would 'deep-six' an incriminating one – that is, get rid of it in a secretive way. The term is supposedly derived from navigation in earlier centuries, where anything more than six fathoms (36 feet) deep under water was unlikely to be recovered. This may also be related to burial at sea, as some sources suggest six fathoms was the legal minimum depth for the body to be immersed.

AT SIXES AND SEVENS

An expression that refers to a person in a state of confusion or chaos. According to Michael Quinion's World Wide Words (www.worldwidewords.org), the phrase, which has been in use in some form since Chaucer, probably derives from the dice game *hasard*, an old and rather complex French precursor of craps, in which players placed bets on numbers they then hoped to roll. Five and six were considered the riskiest (although in probability terms they'd be no less likely than any other two numbers) and the term was used to describe someone considered foolish or confused enough to bet on them. As the Middle French words for five and six (*cinq* and *six*) entered the English language, they were apparently mistranslated to six and seven, despite the obvious absence of a number seven on the standard die. (It probably helps that 6+7 is unlucky 13 (see p.100).)

SEVENTH HEAVEN

To be in seventh heaven is to be at the very height of ecstasy, similar to 'on cloud nine' (see opposite). The phrase seems to come from religious ideas (in Judaism and Islam, as well as sometimes in Christianity) of heaven as being subdivided into seven levels (sometimes shown as concentric circles surrounding the world), with the seventh (known as *Abja'* in Islam and *Arabot* in Judaism) being the highest one, where God has his throne.

ONE OVER THE EIGHT

This expression, which refers to one who's just over the threshold of drunkenness, seems to come from British military slang in the early twentieth century, where eight pints of beer was apparently considered the amount a man could drink before being drunk. Given that beer was weaker then, and that soldiers tend to have high alcohol tolerances anyway, literally drinking 'one over the eight' now would be inadvisable. Appropriately, though, binge drinking is now officially defined as any drinking session in which more than eight units of alcohol are consumed (for men, anyway – it's six for women).

CLOUD NINE

To be on cloud nine is to be elated or extremely happy – usually as a result of love, substance abuse, or some combination of the two. It has been suggested that the term derives from meteorological classifications of cloud formations, with nine being the highest or the tallest, but most cloud classifications have actually used different numbers, while some older versions of the expression used seven or eight.

THE WHOLE NINE YARDS

In expending all one's effort or resources, one might be said to 'give it the whole nine yards'. It's very difficult to be certain about the origins of this expression, and many (mostly incorrect) claims have been made about it. One of the more credible theories is that American bombers during the Second World War were issued with 27-foot (9-yard) belts of machine-gun ammunition (around 900 bullets), and might have to fire them all on a particularly difficult mission. This seems slightly unlikely, however, given that 'the whole nine yards' doesn't seem to have become widespread until the 1960s. Perhaps a more likely possibility is that the phrase derives from the average capacity of cement-mixing lorries – about nine cubic yards … or a rather rude story about a Scotsman's kilt.

DRESSED TO THE NINES

Like 'the whole nine yards' (see p.19), the origins of this phrase, referring to someone dressed very smartly, are murky. A popular theory ascribes it to the British Army's smartly dressed 99th Lanarkshire Regiment, but that regiment was raised well after Robert Burns used the phrase in the 1790s. As with 'the whole nine yards', it has also been suggested that it refers to nine yards of cloth being required to make a man's suit. If this were true, anyone truly 'dressed to the nines' would also be morbidly obese. *Brewer's Dictionary of Phrase and Fable* suggests a corruption of 'dressed to the eyne [eyes]' or it may simply be because nine is the highest single-digit number.

POLICE RADIO CODES

In any emergency situation, communications failures can be fatal, and managing police radio traffic across an entire city is exceptionally difficult and confusing. The situation was far worse, however, in the 1920s and 1930s, when the very idea of police radio was a new one. Operators, often not comprehensively trained, had difficulty understanding what was said and handling large amounts of traffic. Charles L. Hopper, a radio officer in Illinois, recognized the need for a more efficient method of communication and invented a system to simplify it with numeric codes.

Hopper proposed the ten-code – comprising the number 10 preceding the relevant number – as an efficient method of relaying information which was harder to mishear. The '10–' part is used so that it doesn't matter if the first syllable isn't heard, allowing time for radio adjustment, and the '10–' is often dropped in speech (e.g. 'I need a 29 on a Marge Gunderson'). The number 10 may have been chosen because Hopper was working for District 10 of the Illinois State Police. In 1940, APCO (then the Association of Police Communications Officers) rolled out the system of numeric codes, which have remained quite similar ever since.

Aside from simplifying the process of communication, the 10-codes have another major advantage – they make it comparatively difficult for criminals or intrusive reporters to work out what's being said. At least, they used to – the proliferation of lists like this one, especially on the Internet, having rather robbed the 10-code of its tactical edge.

British police have no real equivalent to the 10-code, opting instead for a charmingly obscene range of abbreviations which unfortunately fall outside our remit here. But American police slang has become entrenched in our popular culture through films and TV, and in some cases (particularly the term '10-4') the adoption of 10-codes by CB radio users.

Some of the more interesting '10-codes'

10-00: Officer down, immediate response needed
10-0: Danger
10-4: Message acknowledged
10-10: Fight in progress
10-14: Suspicious person or prowler
10-16: Domestic disturbance
10-20: Location is ... (One might ask 'what's your 20?')
10-24: Emergency backup needed
10-26: Detaining suspect now
10-29: Check past arrests and outstanding warrants
10-31: Crime in progress
10-32: Person with gun
10-35: Major crime alert
10-39: Unit is to use lights and siren (also referred to as Code 3, see below)
10-40: Unit is to move silently (to avoid alerting suspects)
10-57: Hit and run
10-80: Pursuit in progress
10-105: Dead on Arrival (DOA)
10-109: Suicide

Police responses are also graded numerically:

Code 1: Respond at earliest convenience
Code 2: Respond quickly (whether lights and/or siren are used depends on the police force)
Code 3: Urgent response – lights and siren

Please note that the codes given above are the definitions used by APCO. They've come to vary significantly across police forces and states, and it is for this reason that the 10-code is now being retired. Operations involving multiple forces, particularly in responding to the 11 September attacks and Hurricane Katrina, highlighted the confusion of different police forces trying to communicate with this 'standard' set of codes. As a result, FEMA (the Federal Emergency Management Agency) now prohibits the use of 10-codes where complex emergencies involve more than one police force.

Radio technology has advanced significantly since 1940, both in signal quality (see Five by Five, p.14) and traffic handling, and it's now nothing like as hard to hear spoken phrases as it was when the 10-codes were first introduced. Nonetheless, many officers mourn the loss of what has become an essential part of police culture, finding it difficult to return to ordinary English. (See also: 419 Scams (p.62), 187 (p.29).)

DIXIE

A popular term for the American Deep South – the states of Virginia, North and South Carolina, Florida, Alabama, Georgia, Mississippi, Tennessee, Arkansas, Texas and Louisiana. These slave-owning states seceded from the Union after the election of Abraham Lincoln, forming the short-lived Confederate States of America and sparking the American Civil War (1861–5). In many ways, the war was concerned less with the moral implications of slavery (even the northern states were still deeply racist) than with major differences that had developed in the economies and societies of the two halves, and Southerners often viewed the 'War of Northern Aggression' as an attempt by the ethnically more diverse, industrial and rapidly urbanizing northern states to destroy their more traditional society.

For many Southerners even today, songs like 'I wish I was in Dixie' recall nostalgic memories of the old Southern way of life. The fact that the song in question depicts a racist caricature of a freed slave who wants to go back to slavery on a plantation has led to considerable controversy surrounding its continued performance in the South, much like the flying of the Confederate flag.

The term may simply be derived from the Mason-Dixon line, which divided free and slave states, but it could also refer to the number ten. In the days before centralized banking and minting, it was fairly common for even relatively minor banks to print their own currency. One bank in the state of

Louisiana, which was named by the French for King Louis XIV and retains a strong francophone tradition today, printed their own ten-dollar bills with the word *dix* (French for 'ten') on one side. These rare bilingual notes were particularly sought after, and the area (and thus the South as a whole) became known as Dixieland, or simply Dixie.

NINETEEN TO THE DOZEN

This term (also sometimes ten or twenty to the dozen) refers to very energetic activity, usually talking extremely fast. Etymological facts are especially thin on the ground here – no one seems to have an explanation for why this term is used.

23, SKIDOO!

This slang expression from the past is still occasionally found now, referring to taking the opportunity to clear off, leave quickly, or being socially or physically compelled to do so. Although it started off well before this time, the phrase took the United States by storm in the 1920s, and even then its origin was disputed. The 'skidoo' is almost certainly a corruption of 'skedaddle', but the source of '23' is less clear.

Perhaps the most amusing theory is that it refers to 23rd Street in New York, where the city's iconic Flatiron Building was constructed in 1902. At a time when skyscrapers were

only in their infancy, the building's odd angular shape led to complex wind tunnelling down 23rd Street, which tended to cause havoc with women's skirts, making them billow in an unseemly manner. This being 1902, the possibility of catching a glimpse of ankle attracted a number of voyeurs, who were moved along by policemen giving them the '23, skidoo'.

However, most sources put the use of 23 (with no 'skidoo') as a slang term a few years before, with reports of annoyed New Yorkers yelling the number at beggars. Some have claimed '23' is telegrapher's Morse code for 'away with you', much as -30- (originally meaning 'end of transmission') is still sometimes used by journalists to signal the end of a story. Although this explanation is rather uncertain, the use of numeric codes by telegraphers certainly helped popularize the use of numbers as slang expressions in this period (see 86, p.28).

A third popular explanation is that '23, skidoo' derives from a highly regarded 1899 theatre production of Charles Dickens' *A Tale of Two Cities.* The climactic final scene, in which the hero Sydney Carton is guillotined, has an old woman counting the victims, and shouting 'twenty-three!' as Carton dies. This melodramatic scene was parodied by comedians of the time, and it almost certainly popularized the expression, even if it isn't the origin.

FORTY-NINERS

With the discovery of gold in the Sierra Nevada in 1848, the California 'Gold Rush' drew hundreds of thousands of 'forty-niners' (as in 1849-ers) to the area, and this went on until about 1855. Although most forty-niners did not find the massive wealth they expected, they generally made some profit, and their presence created extremely fast economic growth in the area. A local government and infrastructure had to be established quickly, and California became a US state in 1850. Even now, California is America's most prosperous, populous and glamorous state.

77

This otherwise unremarkable number took on enormous importance for Sweden during the Second World War. Seventy-seven, written *sjuttiosju* in Swedish, is extremely difficult to pronounce in that language, and it was the pronunciation – or mispronunciation – of it that helped guards at the border of neutral Sweden to distinguish between native Swedes and others from Germany or occupied Norway.

86

This expression, dating back to jazz-age America, and meaning to 'dismiss' or 'get rid of', has its roots in American restaurants and diners, and originally referred either to being out of stock of an item on the menu, or to ejecting a troublesome client. As with '23, skidoo' (see p.25), numerous largely unsupported theories have been offered. Supposedly, section 86 of the New York State liquor code defined the circumstances under which a patron could be ejected (i.e. '86ed'), but this seems unlikely, given the origins of the expression. It could also be rhyming slang for 'nix', but if so it would be an unusual instance of American rhyming slang.

187

This number refers to section 187 of the California Penal Code, which defines the criminal act of homicide, and '187' has become an American slang term for murder, particularly in major cities. The term appears to have spread to gangs across America, as well as to more general usage, through West Coast 'gangsta rap' in the 1990s, with a number of songs making reference to 187s. A 1997 film, *One Eight Seven*, about gang violence in an LA high school, further popularized the term. The term '419 scam' (see p.62) is derived in a similar way.

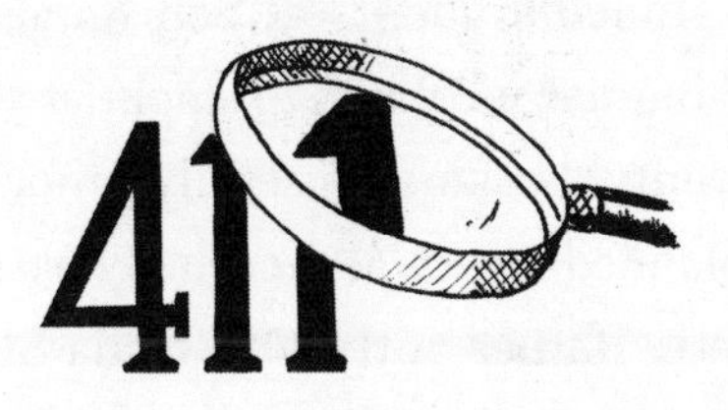

411

The telephone number used for directory assistance in most parts of the US and Canada, '411' is American slang for essential information. It's pronounced – like 911 – as three separate digits, and always used as a noun, similar to 'the lowdown' – one would 'get the 411' on someone or something rather than '411ing' it.

420

Because cannabis use is illegal but comparatively mainstream, the number '420' is used, mainly in the US, as a code for cannabis use in everyday situations. For example, particularly liberal-minded students might advertise shared flats as being '420 friendly'.

Like '187' (p.29), '420' is a three-digit code for illicit activity, which might lead one to think it refers to a penal code or that it's related to 419 scams (p.62) in some way, being the next number along. In fact, though, the origins of this widely used term are surprisingly esoteric. In the early 1970s, a group of legendarily stoned students at San Rafael High School, California, held regular meetings to smoke at 4:20 pm, and the phrase 'four-twenty' became a usefully innocuous-sounding reminder of a planned meeting, eventually passing into heavy usage from there. Rather aptly, California State Senate Bill 420, passed in 2003, clarifies the politically liberal state's law on medical use of the drug.

NUMBERS IN FICTION

00000

The mysterious number at the heart of *Gravity's Rainbow*, a fascinating but exceptionally confusing novel by Thomas Pynchon, published in 1973. The book touches on religion, maths, quantum physics, conspiracy theories and obscure 1940s pop culture as it follows a bewildering cast of characters in a multi-stranded epic storyline centering on an elusive V2 rocket (numbered 00000), in the closing years of the Second World War.

π: FAITH IN CHAOS

Darren Aronofsky's inventive 1998 debut film stars Sean Gullette as Max Cohen, a reclusive mathematical genius with a pervasively numeric worldview, as he tries to uncover mathematical patterns in the world around him. Cohen is fascinated by the digits of π (see p.129), but the film also touches on the Golden Spiral (see p.47) and Kabbalistic gematria (see p.79). It's one of my favourite films, and Cohen's tendency to see numbers in everything partly inspired this book.

THE PRISONER

A highly popular 1960s British TV series starring Patrick McGoohan, *The Prisoner* followed on from the spy series *Danger Man.* The titular incarceree is a former secret agent who resigns but is confined by his former masters to a surreal village until he gives them the information they want. The prisoner is frustrated at being known simply as Number Six, prompting his famous cry 'I am not a number. I am a free man!'

SE7EN

David Fincher's grim, gory neo-noir 1995 film stars Morgan Freeman and Brad Pitt as detectives on the trail of a fanatically religious serial killer (Kevin Spacey) who, in various unpleasantly apt ways, kills those he judges to be committing the Seven Deadly Sins (see p.70), starting with an obese man killed by extreme force-feeding.

THE SEVEN SAMURAI

A seminal Japanese film about ... seven samurai (technically *ronin*, see p.40), who are brought together to defend a village against marauding bandits. The 1954 film's treatment of the effects of violence was extremely influential across the world and across genres, while the plot device of several very different heroes brought together to fight a common enemy was particularly innovative – as well as appearing in the film's Western remake *The Magnificent Seven*, it's the basis for *Ocean's 11*, *The Dirty Dozen* and countless others.

007

The number assigned, of course, to James Bond, the British secret agent created by Ian Fleming who has since appeared in innumerable books and films, though the character (who, for a spy, is remarkably eager to introduce himself to his enemies) owes far more to fantasy than reality. Matt Damon's description of Bond as an 'imperialist, misogynist sociopath' doesn't always seem far off the mark, either. The '00' indicates that Bond is one of few agents with a licence to kill; the '7' appears to have been chosen simply for its mystical and lucky connotations (see p.70).

8 MILE

A 2002 film starring American rap star Eminem (and apparently partly based on his life), which is about a young white working-class rapper in Detroit who tries to become successful in this predominantly black subculture, and who is confronted by various racial and class divisions as he does so.

The film is named after the 8 Mile Road, one of a number of major roads that make up a grid (with a mile between each road) in the area around Detroit. The road has long been known as a socio-economic dividing line between the rich suburbs to the north and the deprived inner city to the south, and therefore serves as a relevant setting for the film.

8½

This 1963 autobiographical film by the Italian director Federico Fellini (1920–93) follows a struggling director who suffers from a creative block in the middle of making a film, and retreats into his memories and dreams. With great simplicity, the title comes from the number of films Fellini had directed before this one – six full-length films and two short films (which he added together to make one full one), plus a film he co-directed (which he counted as half a film). The total is 7½ , making *8½* his 8½th film.

UP TO 11

A classic line in the seminal film *This Is Spinal Tap* (1985), in a scene where guitarist Nigel Tufnel (Christopher Guest) is showing off his guitars and amplifiers. Tufnel demonstrates a special Marshall amplifier with an extra volume setting, in case he needs to go 'one louder'. The phrase 'up to eleven' has come to refer simply to taking things to slightly absurd extremes, and real amplifiers with volume dials going 'up to 11' have been produced.

12 MONKEYS

A superbly innovative science-fiction/mystery film, directed by former Monty Python member Terry Gilliam in 1995, *12 Monkeys* stars Bruce Willis as an inmate of a dystopian prison in a post-apocalyptic future, who is sent back in time (with the aid of an unusually imperfect time-travel process) to stop the disaster. Back in the 1990s, he meets a mental patient (Brad Pitt), whose involvement with the environmentalist extremist group the Army of the Twelve Monkeys seems to be the answer. The title may be suggested by the appearance of twelve monkeys in L. Frank Baum's *The Magic of Oz*.

21 GRAMS

A 2003 film directed by Alejandro González Iñárritu and starring Benicio del Toro, Naomi Watts and Sean Penn. The film, which follows interweaving plotlines around a fatal car accident, is named for the supposed weight of the human soul (about 0.75 oz), a figure that has more weight as a religious and cultural idea than it does as any sort of scientific datum.

The number is derived from a series of experiments performed by Dr Duncan MacDougall in Massachusetts in 1907, which aimed to measure a loss of weight at the moment of death, and thus 'prove' that the soul existed and had physical mass. Unsurprisingly, MacDougall's experiments were deeply unscientific in their methods and ultimately inconclusive. Not only did he record a fairly wide range of weights, some of which increased over time, but his sample size (see p.145) only included six patients. Some subjects in other experiments *gained* a tiny amount of weight at death. Furthermore, such small losses of weight at death can easily be accounted for by the lungs emptying, by the evaporation of fluids, and by simple experimental error.

CATCH-22

Coined in the novel of the same name by Joseph Heller, the term 'Catch-22' has come to mean a no-win situation created by circular logic. The book is a vicious, hilarious, dark satire of bureaucracy, the military and society in general, set among the USAAF pilots of the Fighting 256th Squadron (often jokingly referred to as 'Two to the Fighting Eighth Power') on an island in the Mediterranean, where they're confronted by bizarre military bureaucracy as well as the horrors of war. Among the pilots, Catch-22 applies to those attempting to avoid flying missions on the grounds of insanity. Catch-22 stipulates that anyone who would fly missions at all must be insane, but that a pilot attempting to claim insanity (and thus avoid flying) is demonstrating self-preservation and hence sanity, and can therefore be sent out on missions.

The choice of 22 is, aptly, a rather complex and convoluted story. In the early stages, Heller had intended to call it Catch-18 (a number significant to Heller's Jewish heritage, and Jewish themes which were prominent in the early drafts – see p.81 for why), but the number had to be changed as the lengthy writing process finally came to an end, in order to avoid confusion with a contemporaneous war novel, *Mila 18* by Leon Uris. The repeated digit in Catch-11 seemed appropriate to the repetition that occurs throughout the book, but was considered too close to the film *Ocean's Eleven* (the 1960 Rat Pack version, of course). Catch-17 and Catch-14 were also rejected, for being respectively too similar to the war film *Stalag 17* and not 'funny' enough, before '22' was settled on.

THE 25TH HOUR

David Benioff's novel, filmed in 2002 by Spike Lee, follows a convicted drug dealer on his last day of freedom before going to prison.

THE THIRTY-NINE STEPS

The classic spy novel by John Buchan about a conspiracy to plunge Europe into war and steal British military secrets. 'The thirty-nine steps' in question refer to a meeting point, but the phrase is also used by the enemy spies in the book as a mysterious sign. The book has been filmed four times, most famously (and with substantial alterations to the plot) by Alfred Hitchcock in 1935.

The novel was written in 1914 while Buchan was staying at Broadstairs, Kent, for health reasons. His daughter, then aged around six and still learning to count, went down a set of stairs leading to the beach, and on returning to her father, proudly announced that there were 39 steps.

42

The meaning of Life, the Universe and Everything, in *The Hitch-Hiker's Guide to the Galaxy* series by Douglas Adams. The enormous supercomputer Deep Thought produces this cryptic answer, requiring the construction of an even larger machine (I won't spoil the surprise) to work out the question. Adams is believed to have chosen the number entirely at random, and not with an actual hidden meaning in mind, though many have tried to find one.

THE 47 RONIN

Probably Japan's best-known folk tale – and apparently one with a fairly solid basis in fact – is that of forty-seven samurai whose master, Asano, assaulted an official, Kira Yoshinaka, and was compelled to commit honour suicide (*seppuku*). Without a master, the samurai became *ronin* – wandering outcasts who acted a little like mercenaries.

The forty-seven were compelled by the samurai code of honour (*bushido*) to avenge Asano, and constructed an elaborate plan to do so. In order to put Kira off his guard, they would deliberately give the impression of being dishonourable – drinking and debauching themselves, and never showing interest in avenging Asano. After two years of this charade, they were able to enter Kira's house easily, and killed him.

The *ronin* had known from the beginning that this course

of action would end in their deaths. They turned themselves in to the authorities (except one, who had departed to carry back the news of Kira's death, and seems to have been spared), and were sentenced to death for murdering Kira. Because they had acted according to the precepts of *bushido*, however, they were allowed to commit *seppuku*, a more honourable death than that of an ordinary murderer.

HAWAII FIVE-O

A long-running American cop show, much loved around the world, set in, and named after, the fiftieth state of the Union.

ROOM 101

The room where dissenters are tortured with their greatest fears in George Orwell's *1984*. Orwell worked for the BBC for some time, and is thought to have numbered the room after a hated conference room on the first floor of BBC Broadcasting House. Such was the cultural influence of the Room 101 in the novel, that the chief of the Stasi, Erich Mielke, had the rooms in his building renumbered so that his office would be Room 101 and, in a wonderfully circular development, the BBC now have a well-known television series called *Room 101*, in which celebrities discuss their pet hates.

LES 400 COUPS

Named after a French expression similar to 'the third degree' (see p.13), *Les 400 Coups* – literally, the 400 blows – is the first in a series of heavily autobiographical films by French New Wave pioneer François Truffaut. The film follows the adventures of a young French boy – Antoine Doinel – as his mischievous ways are harshly judged by society.

FAHRENHEIT 451

Ray Bradbury's bleak novel, about an anti-intellectual dystopian future in which all books are burned, is named for the temperature at which paper burns (about 233°C).

24601

Valjean's prisoner number in *Les Misérables*, since used for many fictional prisoners. A questionable popular claim is that it was the date of Hugo's conception – 24 June 1801 – but if true this would make his birth (in February 1802) a premature one.

NUMBERS IN CULTURE

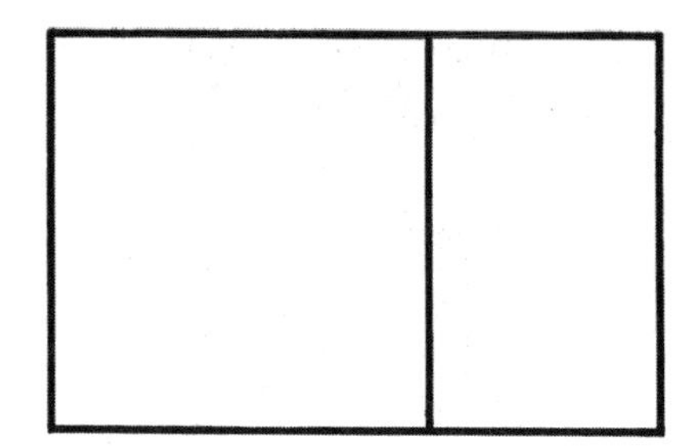

1.618 – THE GOLDEN NUMBER

Although not commonly known today, the Golden Number or Golden Ratio (1.6180339887..., also known simply as φ (phi)) has been credited with great importance in art and design for over two thousand years, as well as being of mathematical interest.

A perfect rectangle?

The Golden Number's main contribution to aesthetics comes in the form of the Golden Rectangle – a rectangle where one side is φ times longer than the other, meaning the Golden Rectangle can always be divided into a square and another Golden Rectangle (see illustration above).

φ may have originated in Ancient Egypt, appearing in the proportions of the pyramid at Giza and in admittedly vague references to a 'sacred ratio'. It was certainly important to the Greeks – the Greek sculptor, painter and architect Phidias used it in his design of the statues in the Parthenon, and it is designated φ after him (the first letter of his name in Greek).

Plato and Euclid later studied the Number, although the term 'Golden Number' only emerged during the Renaissance. The

supposed significance of the Golden Number in Greek art and architecture (particularly the claim that it formed the basis for temple front designs) appears to have been exaggerated, but the renewal of interest in classical disciplines during the Renaissance brought enormous interest in the Golden Number and the Rectangle, among other theories of anatomical and aesthetic proportion. In particular, Leonardo da Vinci's illustrations for Luca Pacioli's *De Divina Proportione* applied the Golden Ratio to human faces, and many claim that his paintings, including the *Mona Lisa*, employ the Golden Ratio very heavily in their composition.

More recently, the twentieth-century painter Piet Mondrian used the number to compose his abstract geometric works, and the pioneering Modernist architect Le Corbusier used the Rectangle to develop his Modulor system of architectural scaling, designed to mimic human proportions.

A degree of scepticism is needed, however, when considering the rectangle's importance in aesthetics and proportion. In particular, it's important to bear in mind that body and facial proportions vary widely between individuals, not always coming especially close to any Golden form. As in nature, many of the sightings of the Golden Rectangle in art and design have been disputed.

Point Six One Eight Zero Three Three Nine Eight Eight Seven ...

Like π (see p.129), φ is an irrational number (see p.121) – a decimal whose digits never end and never repeat. Mathematically, the Golden Number is defined by the ratio of two different positive constants, a and b, a being the larger of the two. If a + b is the same, proportionally, to a as a is to b, a will be φ times larger than b (and a+b φ times larger than a), i.e.:

$$\frac{(a+b)}{a} = \frac{a}{b} = \varphi$$

φ is also equal to $(1+\sqrt{5})/2$.

φ bears a close mathematical relationship to the Fibonacci sequence (see p.140). Look what happens if each term of the sequence is divided by the previous one:

1÷1 = 1
2÷1 = 2
3÷2 = 1.5
5÷3 = 1.666666 ...
8÷5 = 1.6
13÷8 = 1.625
21÷13 = 1.61538462
34÷21 = 1.61904762
55÷34 = 1.61764706
89÷55 = 1.61818181 ...

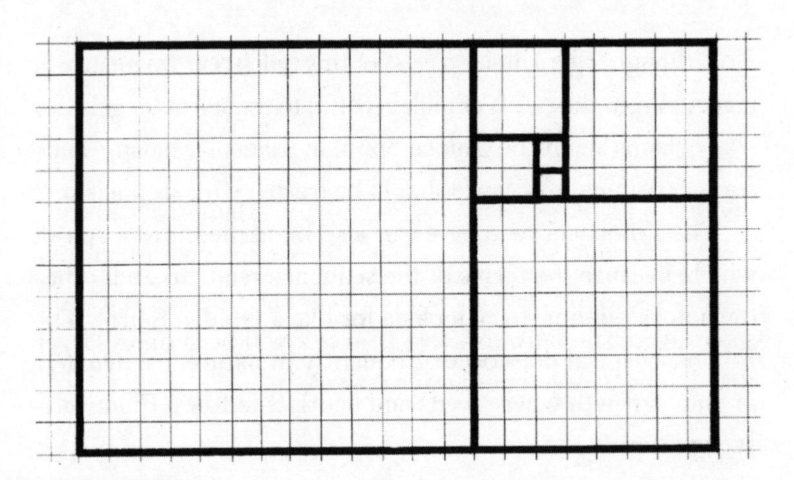

The relationship extends to geometry as well. If the Fibonacci sequence is represented as a series of squares (1×1, 1×1, 2×2, 3×3, 5×5, etc...), they rapidly start looking like a Golden Rectangle (see illustration above).

The corners of the Golden Rectangle can be joined to form a Golden Spiral, which gets 1.6180339887 times wider with each 90-degree turn:

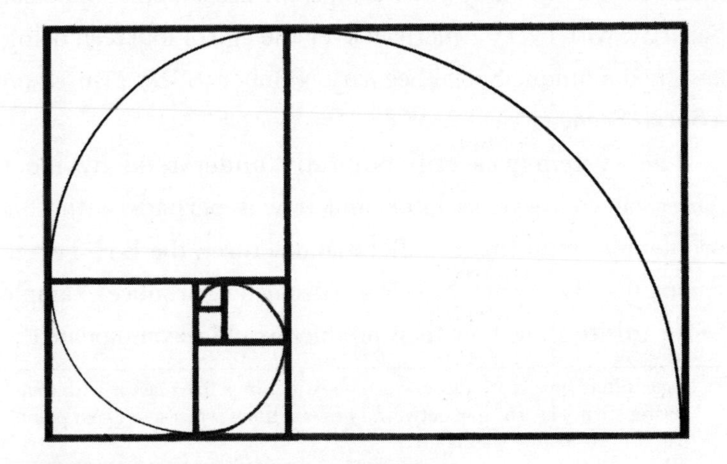

Although logarithmic spirals* of this type occur frequently in nature – from the swirls of milk in coffee to the spirals of galaxies – any claims about the Golden Spiral in particular having some sort of cosmic significance should be treated with scepticism.

The Fibonacci rectangle can also be turned into a spiral, which, because the terms of the sequence relate to each other by factors tending to φ, looks a lot like a Golden Spiral. The Fibonacci spiral does occur frequently in nature, particularly in plant growth where seeds and fruitlets follow a Fibonacci-type pattern.

078-05-1120

In 1936, America was still recovering from the Great Depression, and Franklin D. Roosevelt's wide-ranging New Deal programme of social spending was being put in place. It included, among many other things, the introduction of Social Security, with every American over the age of fourteen being assigned a unique Social Security Number (SSN), printed on a Social Security card.

The system was still not fully understood by most Americans two years later, and it was perhaps with this problem in mind that a wallet manufacturer, the E.H. Ferree company of Lockport, New York, decided to produce example SSN cards to show how their product would accommodate the

* Logarithmic spirals are those which get wider by a fixed factor with each quarter turn – i.e. the gap between lines is x times wider at a given point than it is 90° before that point.

new cards. The example card closely resembled a real one, despite the word 'SPECIMEN' stamped on it, and the SSN used – 078-05-1120 – was a valid one taken from Mrs Hilda Whitcher, a secretary at the company.

When the wallets were distributed to shops nationwide, thousands of confused buyers decided that SSNs must be being issued through wallet purchases. In the peak year 1943, nearly six thousand people claimed the number as their own (the unfortunate Mrs Whitcher, by this point, was not among them, having been issued a new one).

Similar problems have occurred with duplication of SSNs, but the abuse of 078-05-1120 was by far the worst. Even as late as 1977, twelve people were still using the number. To prevent this kind of thing happening again, the Social Security Administration assigned a block of invalid SSNs (987-65-4320 to 987-65-4329) solely for use in advertising (similar to the 555 area code, see p.56).

THE INDIANA PI BILL

In 1897 came one of the strangest incidents in mathematical history – the Indiana Pi Bill. It was proposed by Dr Edwin J. Goodwin, an amateur mathematician who was obsessed with squaring the circle – an ancient mathematical problem that had troubled mathematicians for centuries, until ultimately proven impossible – and almost got his crank theories accepted by the state's senate.

The discredited theory held that it was possible to find a circle's area by using a straight-edge and a set of compasses to construct a square of the same area as the circle, and by finding the area of the square, one could also find the area of the original circle. The problem was that this was reliant on π (see p.129) not being a transcendental number, but in 1882 Ferdinand von Lindemann had proved that it was, and hence that squaring the circle was impossible. Goodwin carried on claiming he had done so, however, listing it alongside other impossible accomplishments in the text of the Bill:

> 'his solutions of the trisection of the angle, duplication of the cube and quadrature of the circle having been already accepted as contributions to science by the *American Mathematical Monthly*, the leading exponent of mathematical thought in this country. And be it remembered that these noted problems had been long since given up by scientific bodies as insolvable mysteries and above man's ability to comprehend.'

Pi in the face

Dr Goodwin's bill proposed, seemingly without irony, a 'new and correct' value for π – 3.2. Crammed with impenetrable language and convoluted, self-contradictory maths, which many senators were unable to understand, he generously allowed the Indiana school system to use his profoundly wrong thesis free of copyright. In fact, the bill would probably have passed were it not for the timely intervention of C.A. Waldo, a mathematics professor at nearby Purdue University. It was Waldo who explained the problem with Goodwin's maths to the senators, who postponed the bill indefinitely, and Goodwin was laughed out of the Senate house.

NUMBERS GAMES

Numbers games are illegal lotteries that were extremely popular among America's urban poor from around the mid-1850s to the 1950s, after which they were largely superseded by state lotteries. Typically, players would choose three or four numbers between 0 and 999, although the range of available numbers, and hence the odds of winning, varied significantly (later on, the operators chose independent numbers such as the US Treasury balances that were published daily in American newspapers). Numbers that were especially likely to be picked – lucky 7, for instance, (see p.70) – were often eliminated from the draw. Attractive because of their cheap tickets and promises of enormous wealth, numbers games were also often fairly convenient, as number 'runners' went from door to door to record players' bets.

Numbers games were also sometimes known as 'policy', a euphemistic suggestion that regular betting on them was an insurance policy against the future, as you were supposedly bound to win one day. I hardly need to add this was not the case – many numbers games were rigged, and the organizers took massive cuts. Police officers who took an interest were often paid off, so very few arrests were made.

4-11-44

Although not invented by black Americans, numbers games were especially popular in black communities and in Harlem, where they even attracted middle-class players. In the north-eastern US, illegal gambling was viewed as being essentially a 'black problem' and the popular 4-11-44 bet became a byword for black poverty, frequently used in racist songs that portrayed

African-Americans as superstitious and irresponsible. The 4-11-44 bet also appears in many blues songs of the time, but its origins are still uncertain.

Numbers games acquired a particular popularity in these communities in the 1920s, and the results were not entirely harmful. The racist mockery of the 1900s and 1910s had given way to a backlash – the Harlem Renaissance – in which African-Americans rediscovered their cultural identity, fostered a sense of pride and developed the doctrine of 'buying black' – encouraging black residents to choose local black businesses (and boycott racist ones) and thus reinvest black dollars in their community. Policy bankers in Harlem got rich from the proceeds of illegal gambling, but they were also among the first to lend money to Harlem's poorest residents.

Although the Great Depression saw an enormous surge of interest in numbers games as desperate people clung to a chance of financial salvation, it also saw a change in the balance of power among numbers operators. The Mob, which had been content to keep out of numbers games, lost a great deal of money to falling sales in other illicit sectors and decided to compensate by taking over Harlem's numbers games with intimidation and violence. The end of Prohibition in 1933 and a mass crackdown on corruption in law enforcement saw a reduction in the kind of large-scale numbers rackets that occurred in Harlem and other poor communities.

Illegal gambling, however, is still very much alive.

WHY BUSES COME IN THREES

The hugely irritating tendency of buses to lag behind advertised timetables, before arriving in a seething glut half an hour too late, is a well-documented phenomenon of public transport known as 'bunching'. Bus companies have invested a lot of money and effort in attempts to alleviate it, and advances in communications technology mean that the worst bunching can now be more easily spotted.

The problem, though, is essentially mathematical in nature, and difficult to eradicate completely. Furthermore, unlike on railways and the underground, where all trains are monitored and can be held in stations to even out the gaps between them, buses have to work within a much larger organic system of traffic, where they may be forced to keep moving.

Let me explain

On a given bus route in a major city, the buses leave the depot at fixed intervals a few minutes apart. As the buses start at 8.15 in the morning, a significant number of morning commuters have turned up at most or all of the stops on the route. The first bus stops at all of them, picking up a large number of morning commuters, which causes it to spend a fairly long time at each stop.

The next bus, departing at 8.25, also stops at every bus stop, but this time only picks up the smaller number of commuters who've turned up in the last 10 minutes, thereby spending less time at each stop, while the first bus, only a few stops ahead, is still picking up lots of people. This second bus begins to catch up with the first one. The third bus gets even closer as the number of commuters waiting on the route decreases further, and so the process continues. After rush hour, it tends to even out a little, though, as there are far fewer people, and buses can be spaced out without making people late.

But because the regularity of buses is not matched by the comparatively unpredictable patterns of passengers turning up at the stops, bunching can be brought on by an unusually large group of people arriving at a single bus stop at any given time, and thus forcing one bus to stop for quite a while as others close the distance to it.

As the bunching process is repeated over the whole day, especially when combined with traffic lights and other road delays which stop the buses in front and allow the ones behind to catch up, buses get bunched up on their route, and by the end of the route large gaps may appear in the bus service, coming between groups of several buses arriving all at once.

555

The problem with phone numbers as plot elements in TV and film is that they tend to exist in reality, usually resulting in thousands of nuisance calls by those with too much free time. American rock band Tommy Tutone's 1982 hit '867-5309/ Jenny' – about a girl's phone number written on a toilet wall – is still remembered for a spate of prank calls to the number, rather than anything else.

American films and television have traditionally avoided such issues by agreeing with telecoms companies on an area code reserved exclusively for fictional purposes. The practice appears to have begun in the 1950s and 1960s, with an early example in the film *A Patch of Blue* (1965), which also established a canonical fake phone number – 555-2368 – since used in *Ghostbusters, Close Encounters of the Third Kind*, and *Memento* amongst others.

It seems likely that 555 was chosen, and gained currency, partly because of its memorability as a repeated digit, but also because (in the days of named telephone exchanges), there were no major English place names that combined the letters J, K and L (the letters assigned to the number 5 on the keypad). The 555 number was sometimes given a fictional exchange name as well, such as Klondike-5 (KL5), which features in phone numbers in *The Simpsons.* The suffix 2368 was probably chosen for its use in 1940s telephone advertising, which referred to Exchange 2368 as a CENTral exchange (look at the letters on the keypad again).

With growing pressure on the telephone system, the administrators of the North American Numbering Plan have opened up the 555 exchange code, and now only a certain range (555-0100 to 555-0199) is reserved for fictional use.

There is no true British equivalent to the 555 number, and the exchange code is used in this country like any other. However, Ofcom has reserved the 01632 area code, blocks of numbers in most major British cities (usually in the 496-0000 to 496-0999 range), and all mobile numbers beginning 07700 900, for fictional purposes.

Ringing the Almighty

Recently, some films and TV series have bucked this trend. *Bruce Almighty* (2003), for example, featured a personal number for God. This turned out to belong to a shop manager in Salford, who received hundreds of calls asking for help and forgiveness. And a few asking for smiting, I would imagine. The television series *24* and *Scrubs*, among others, have been shrewder, incorporating real phone numbers as promotional devices – the number for CTU actually belongs to the *24* production staff, while Scrubs' Dr Chris Turk is assigned the number 1-916-225-5887 – 1-916-CALL-TUR (k) – which connects to an answering machine and a message from cast members.

NUMBERS STATIONS

These cryptic radio transmissions are broadcast all over the world, usually on shortwave bands, and consist mainly of long spoken lists of numbers. The very strangeness of numbers stations has attracted a subculture of enthusiastic listeners.

Most people agree that numbers stations are linked to espionage, and this was all-but-officially confirmed by a DTI spokesman in 1998, who added that 'they are not for, shall we say, public consumption'. It seems likely that the broadcasts can only be decoded by an agent with the correct 'one-time pad' – a list of randomly generated numbers that is used to encode and decode a single message, before being destroyed – as anything else would be rather unsafe.

Although it's fairly easy to trace the origin of radio broadcasts, numbers stations usually broadcast from many hundreds of miles away, and are usually located on territory friendly to their agents. It's difficult to trace radio listeners and almost impossible to tell what's being listened to, and numbers stations have survived for decades as a safe method of communication. Phones and internet connections can be (and often are) automatically monitored for suspicious phrases, but tracing one spy amid the millions of ordinary radio listeners in any city would be absurdly impractical and probably not worth the expense. The broadcasts usually follow a regular schedule, which might be broken in exceptional circumstances (such as a Moscow broadcast during the 1991 Soviet coup which consisted of constant repetitions of the number five). Each broadcast is usually introduced by a short, distinctive audio clip, presumably to make it easier for agents to tune in – MI6 used two bars of 'The Lincolnshire Poacher' for many years. After this comes a list of numbers read by an automated voice, usually repeated once to avoid errors.

There has only been one criminal case involving a numbers station – a Cuban broadcast known as '¡Atención!' – in the mid-1990s. Infamously badly run and plagued with technical problems, ¡Atención! had conclusively revealed its origin by accidentally rebroadcasting Radio Havana on its frequency. ¡Atención! was broadcasting to the Wasp Network, a group of five Cuban agents in Florida who used a computer program to decipher the numbers. When FBI agents broke into the Cubans' flat and copied the decryption program, they discovered a plan to infiltrate anti-Castro groups, which led to the 'Cuban Five' being sentenced to long prison terms. Their case remains controversial, as the group they were infiltrating is considered a terrorist group in Cuba and the information they sent back was not classified. Many politicans and human rights groups have petitioned the United States to release them.

THE 23 ENIGMA

This numerological theory claims that the number 23 recurs in many different and surprising places. Supposedly, 23 can be connected to any significant event or phenomenon, good or bad. For instance, people have 23 pairs of chromosomes, Aleister Crowley (who started studying magic at the age of 23) defined 23 as 'the number of life', and the Earth's axis is tilted at an angle of 23.5°. The 23 Enigma (which, by the way, is unrelated to the old slang expression '23, skidoo!') has inspired two films, *23* (of which more below) and *The Number 23*, a widely panned Jim Carrey effort about a man who becomes obsessed with the 23 Enigma. Carrey even named his production company JC23, not that that did the film much good.

The idea that 23 could somehow have mystic power because it can be linked coincidentally to certain events is, of course, deeply irrational and entirely unconvincing. In this case, however, that's pretty much the point. The 23 Enigma originates not in any ancient text, but in a series of novels published in 1975 called the *Illuminatus!* trilogy, by Robert

Shea and Robert Anton Wilson. The trilogy is a complex satirical conspiracy story, and numerology, including the significance of the number 23, is one of its major themes.

Wilson appears to have been inspired by a story he heard from the author William S. Burroughs. While in Tangier in 1960, Burroughs made the acquaintance of a sailor, Captain Clark, who boasted to Burroughs that he had run his ferry service for 23 years with not a single accident. That same day, the ship sank, killing Clark and all the passengers. Burroughs was ruminating on this news that evening when he heard a report on the radio of a plane crash in Florida. The plane was Flight 23, piloted by a Captain Clark.

It's not funny ...

Wilson and the *Illuminatus!* trilogy are strongly linked to Discordianism, a postmodern 'joke' religion*, which celebrates confusion and chaos and claims to worship Eris, the ancient Greek goddess of discord. The 23 Enigma seems intended as a joke about the self-fulfilling nature of numerology and the ease with which human pattern recognition can be fooled. The numbers 17 and 5 are similarly revered by Discordians, and can be said to be related to 23 because the digits of 17 add up to 8, or 2^3, and of course $2 + 3 = 5$.

It wasn't so funny, however, for a German hacker named

* The postmodernity of the joke here is apparently derived from blurred boundaries between what's meant to be serious and what isn't. The 'holy book', *Principia Discordia*, ends with the words, 'If you think the *Principia* is just a ha-ha, then go read it again.' As Moe says in *The Simpsons*: 'It's PoMo ... Postmodern? Yeah, all right, weird for the sake of weird.'

Karl Koch. Koch and his associates hacked into a number of American military systems from Hanover and Berlin and sold the data to the KGB. Koch was obsessed with the *Illuminatus!* trilogy, using the name of the book's protagonist, Hagbard Celine, as a pseudonym for his activities. He was also severely paranoid and addicted to cocaine, and became convinced that he, like his fictional namesake, was fighting against a vast Illuminati conspiracy. Koch was captured, along with those he was working with, in March 1989, provoking an international spy scandal. He eventually killed himself on 23 May 1989 – at the age of 23. The 1998 film *23* is based on his life.

419 SCAMS

You've probably received the odd email from 419 scammers. Typically from Nigeria or elsewhere in Africa, these missives appeal in broken English to naivety and greed. You know the story: there's an enormous sum of money stuck in a far-flung bank account that'll be forfeit if it's not removed from the country soon, and YOU, as a foreigner, have been selected to help. You're promised a sizeable cut of the stricken money in exchange for your help, but you do need to pay a handling fee first, for which they'll just need your bank details ...

Several further techniques are used to add an air of authenticity or relatability. In many cases, the money has been left behind after the death of a major public figure and the emailer is a grief-stricken friend or relative, earning the sympathy of the victim. Particularly religious recipients can be swayed by frequent reference to Jesus. Elements of truth – an air crash death that actually occurred, for example – can be used to give a shred of credibility to the story.

This form of advance fee fraud is known as a 419 scam, after Section 419 of the Nigerian penal code, which concerns confidence tricks of this type (in much the same way as 187s, see p.29). In recent years the poor English and obvious lies in the emails have made them a source of humour, and Nigerian email scams are now so widespread that many legitimate emails from Nigerian businesses are filtered out by overzealous software.

Known as 'yahoo boys' in Nigeria, 419 scammers are young men who spend hours in Internet cafés, trawling for Western email addresses and paying off any police who might take an interest. In a country still gripped by political violence, corruption and economic uncertainty, there are many worse ways to make a living. Many are able to live the high life on the proceeds of Internet crime, but the 'yahoo millionaire' lifestyle is increasingly condemned as dishonest, superficial and foolish (hence the name, unaffiliated to the search engine Yahoo!, or, as far as I know, to Swift's *Gulliver's Travels*) and many Nigerians are angry at the tarnishing of the country's image abroad.

Yahoo boys are not usually acting out of desperation, nor are they chancers trying it on with email users. Many are affiliated with dangerous organized crime gangs in Nigeria and elsewhere, and the consequences of falling for 419 scams can occasionally be disastrous. The arrangement often involves the victim travelling to Nigeria – a crime if done without a visa, which can then be used by the fraudsters to blackmail their victim. If a victim then tries to get their money back, the gangsters may threaten physical violence, and several people have been kidnapped and even killed when attempting to escape 419 scams.

THE PIRAHÃ TRIBE

For one small Amazonian tribe, few of the numbers discussed here will have even the slightest meaning or relevance. (Otherwise, they'd obviously be rushing down to the local Amazon bookshop for a copy ...)

The Pirahã people (the name is pronounced with the stress on the last syllable) live on the banks of the Maici River in north-western Brazil, and currently number only a few hundred. They have had almost no contact with the wider world since they were found by Brazilian explorers in the eighteenth century. Their culture as a whole is basic and rooted in the immediate present, with no history, no art, and no preparation for the future. Studies of it have been controversial in academic circles, particularly in the world of linguistics.

The Pirahã language is extremely simple, and this extends to numbers, which for the Pirahã are only three – one, two and many. Even these are fairly inexact – translating more like 'one or two', 'a couple', 'a lot'. While this in itself is not unheard-of among remote tribes, the issue here seems to be a deep-rooted cultural one. Most remote tribes with one-two-many counting systems can be taught to count, but the Pirahã seem incapable of grasping the idea. Daniel Everett, a linguist, former missionary and leading expert on the tribe, conducted tests in which he placed a number of objects in front of Pirahã members, and asked them to take the same number from a can and place them in front of him. For numbers below three, the Pirahã did well in the tests, but usually failed when higher numbers were introduced.

In order to preserve their unique culture, the Pirahã tribe's land is now a reservation, but they are under constant threat, their population having declined massively since they were first discovered.

THE 10 PER CENT MYTH

'What's holding you back? Just one fact – one scientific fact. That is all. Because, as Science says, you are using only one-tenth of your real brain-power!'
– advertisement for the Pelman Institute

'Thank God, nurse! The bullet passed through the 90 per cent he didn't use!'
– something no one has *ever* said

You've almost certainly heard this one trotted out at some point – we only use 10 per cent (or another small fraction) of our brains, and presumably would gain amazing intellectual powers if we could only unlock the other 90 per cent. The idea is untrue, and scientifically invalid: it would make no sense for humans to evolve large brains over millions of years only to leave them mostly inactive.

It does, however, seem to have its origins in mainstream science, particularly phrenology, a Victorian pseudoscience which held that brain structure, and hence personality, could be determined by measuring the shape of the skull. The central idea of phrenology – that the brain can be divided up into discrete sections with various responsibilities – has been discredited by modern brain scans. Before these scans, scientists relied on using electrodes to stimulate areas of the brain and see what happened. Areas that produced no obvious physical reaction (like twitching limbs) when stimulated were considered inactive.

Another possible origin is misinterpretation of glial cells (which 'glue' neurons together, outnumbering them about 10 to 1). Additionally, a 1930s experiment by K.S. Lashley on rats found that they could lose most of their brains while still being able to run around.

Possibly the earliest, and probably the best-known, appearance of the 10 per cent myth was in a 1944 advertisement for the Pelman Institute, a self-help correspondence course with rather optimistic claims. 'Pelmanism' was one of the most popular miracle self-help courses in a period full of them, and millions grew up 'knowing' they only used 10 per cent of their brains.

The false statistic continued to appear in adverts for many years, especially those promoting self-improvement and self-help, reaching something of a peak during the self-help boom of the 1990s. With the respect for scientific methodology that has long distinguished them, the psychic lobby, including 'paranormalist' Uri Geller, leapt on the claim as evidence of psychic powers, the idea being that these powers could be discovered if only the unused portion were unlocked.

This idea is a classic example of the argument from ignorance (because science doesn't know everything about the brain, it can't disprove psychic powers, therefore they must exist), but there's a grain of truth in it. Most people do underuse their brains, and the 10 per cent myth plays on ubiquitous insecurity about intelligence and mental powers, which drives the sales of so many modern-day Pelmanist quick fixes.

NUMBERS IN MYTHOLOGY AND RELIGION

SEVEN

The number seven crops up everywhere, particularly when religious or mystical elements are involved. According to Genesis (and some increasingly vocal fringe religious types), the world was created in seven days (OK, six days' work and a day of rest).

The Catholic Church features Seven Sacraments, beginning with three sacraments of initiation into the Church: Baptism (as a child or an adult), Confirmation and the Eucharist (including Mass and Communion). There are also two Sacraments of Healing – Reconciliation or Penance (including confession), and the Anointing of the Sick by a priest; and two mutually exclusive Sacraments of Vocation: Marriage and Holy Orders. A number of these sacraments make up the Last Rites administered to dying Catholics – a final Penance (depending on whether the person can still speak), Anointment and a last Eucharist. In deathbed conversions, hasty Baptism and Confirmation is also necessary.

Christianity also features seven Deadly Sins, which are sometimes shown counterbalanced by seven Heavenly Virtues.

Deadly Sins	Heavenly Virtues
Gluttony	Temperance
Envy	Kindness
Greed	Charity
Pride	Humility
Lust	Chastity
Sloth	Zeal
Wrath	Meekness

Multiples of seven also appear in this way – there are fourteen Stations of the Cross (observed mainly in the Catholic Church), fourteen Holy Helpers (saints venerated in less scientific times as effective against various illnesses), and events in the Bible are often described as happening on the fourteenth day of a given month.

Unsurprisingly given the common origin of the Abrahamic religions, a similar concept of seven divine traits exists in Judaism, represented in the seven branches of the menorah. The Star of David has seven parts (six points and the centre), and there are seven holidays in the Jewish year. In Islam, there are seven verses in the first chapter (sura) of the Qur'an.

The divinity of the number seven may also be the reason for the number six being traditionally unlucky in religion: if seven represents Godly perfection, six represents the imperfections and flaws of man.

Seven's everywhere

The exact reason for seven's significance is not very clear. It could be to do with the seven 'heavenly bodies' visible to the naked eye in the ages before astronomy – the sun, the moon,

Mercury, Venus, Mars, Saturn and Jupiter. This is probably the origin of the Jewish and Muslim (and to some degree Christian) idea of seven heavens (see p.18). It certainly helps that seven is relatively small and a prime number, though.

This theory does at least explain the fact that seven is significant well outside the context of the Abrahamic religions. In Japanese mythology, for instance, there are seven gods of good fortune (*shichi fukujin*), people are reincarnated seven times in Japanese Buddhism, and there are seven principles in the *bushido* samurai code.

We also generally refer to seven colours in a rainbow, even though we rarely see more than four and they don't generally have distinct boundaries. Isaac Newton, who discovered through experiments with prisms that white light was made up of coloured components, added the colours orange and indigo so as to conform to his theory that the colours of the spectrum corresponded to the seven major notes of a musical scale.

The choice of a seven-day week is probably largely to do with convenient subdivision of time. The Babylonians, who laid the foundations of our current measurement of time 5,000 years ago, were the first to divide the year into months based on the lunar cycle*. The lunar cycle is a rather awkward 29 days long, though, so for the sake of convenience the year was divided up into months of 28 days, which could then be subdivided neatly into four seven-day weeks.

Seven is also often used for more worldly collections of things, seemingly to exploit its mythological resonance. 'Seven Seas' referred to a number of different collections of bodies of water, and is still sometimes used today to refer to the Antarctic, Arctic, North and South Atlantic, North and South Pacific and Indian Oceans. It seems likely that it was simply a poetic turn of phrase meaning all the oceans, and there is some evidence for the number seven being synonymous for a time with 'several'.

* Before the adoption of the Julian calendar in England, the country used a lunar calendar, with thirteen months of twenty-eight days. A quick check with a calculator will show that this leaves us one day short of a full year. The old tradition of 'a year and a day' as a standard time period for legal purposes, and in neopagan religions and secret societies probably originated here, with the extra day being to bring the 13 months up to a full solar year.

THE SEVEN WONDERS OF THE ANCIENT WORLD

The Great Pyramid of Khufu
Completed 2560 BC at Giza, Egypt.
It is the only Wonder still standing.

The Hanging Gardens of Babylon
Completed early 6th century BC on the Euphrates, near modern Baghdad.
Destroyed by earthquake in the first century AD.

The Temple of Artemis
Completed c. 550 BC at Ephesus, near modern Izmir, Turkey.
Destroyed and rebuilt repeatedly – finally torn down by a mob in AD 401.

The Statue of Zeus
Completed c. 450 BC at Olympia, Greece.
Probably destroyed by fire in AD 462.

The Mausoleum (tomb of King Mausollos,
hence the name)
Completed 350 BC at Halicarnassus
(now Bodrum), Turkey.
Destroyed by earthquake and subsequent Crusader
disassembly in late 15th century AD.

The Colossus (statue of Helios)
Completed 282 BC at Rhodes, Greece.
Destroyed by earthquake in 226 BC.

The Lighthouse of Alexandria
Completed c. 280 BC in Alexandria, Egypt.
Destroyed by successive earthquakes in
the 14th century AD.

THE NUMBER OF THE BEAST?

> 'Here is wisdom. Let him that hath understanding count the number of the beast: for it is the number of a man; and his number is six hundred threescore and six.'
>
> – **Revelations 13: 18**

666 has long been considered unlucky in Christian traditions because of this passage from Revelations, which associates the number with the Devil. Like thirteen (see p.100), there is a widespread superstition about 666, so widespread that it's catchily known as hexakosioihexekontahexaphobia. It is interesting to note that 2006 brought expressions of concern from pregnant women expecting children on 6 June of that year – not helped by the heavily publicized release of a remake of the film *The Omen* on that date.

Highway 666 in the western US was renumbered to 491 after suggestions that the high rate of accidents was due to Satanic influence, although its safety record probably had more to do with the removal of all the road signs by thieves who sold them on eBay. The town of Reeves, Louisiana, successfully petitioned to change its dialling code from 666 to 749. One wonders whether New York or Los Angeles residents would have been quite as concerned. In China, however, 666 is prominently displayed in shop windows (see Chinese Lucky and Unlucky Numbers on p.88).

There are various theories about why 666 was chosen. Many revolve around the mystic practice of 'gematria', with 666 possibly representing Nero, who was known for his persecution of Christians. See the entry on gematria (p.79) for details on how the numbers are assigned here.

Gematria interpretation

If the Greek phrase Νερων Καισαρ (Neron Kaisar = Emperor Nero) is transliterated into Hebrew, we are left with:

N	R	W	N	Q	S	R
נ	ר	ו	ן	ק	ס	ר
50 +	200 +	6 +	50 +	100 +	60 +	200 = 666

(NB: Hebrew is actually written right-to-left, but is shown the other way round here for clarity's sake.)

Although this can be done with a number of Roman emperors, Nero is perhaps more likely than most to be the Beast, as he also satisfies some of the Greek manuscripts that give the Number of the Beast as 616, if his Latin name, Nero Caesar, is transliterated into Hebrew:

N	R	W	Q	S	R
נ	ר	ו	ק	ס	ר
50 +	200 +	6 +	100 +	60 +	200 = 616

Nero is the only one to satisfy both theories in this way, but numerology, even more than most pseudosciences, is full of vague definitions, suspect methods and absurd gaps in logic. Many cranks over the centuries have tried to apply the Number of the Beast to political or ideological opponents, including Martin Luther, the Pope, Napoleon Bonaparte and most recently Barack Obama.

A more mundane explanation for the origin of 666 is that it's simply used as a general large number. In Roman numerals (see p.137), 666 is DCLXVI – all the Roman numerals below M used once, possibly suggesting simply that the Beast is a large or powerful force. It's also possible that, given the holiness of 7 in the Bible, 6 is intended to represent the imperfection of man (described as being created on the 6th day), and 666 – 'the number of a man' – is simply a numerical representation of the Beast as the imperfections of humanity. It's much easier to blame Satan than oneself, of course, and that's been a favourite activity of religious extremists for hundreds of years.

GEMATRIA AND THE BIBLE CODE

'Gematria' is a term encompassing various long-running traditions of numerology that involve assigning numeric values to letters of an alphabet in order to find some hidden meaning in them. There are a vast number of systems for doing this in the Greek, Hebrew and Arabic alphabets among others, and it's also used in modern numerology (see p.82). The traditional Jewish system of gematria assigns letters in the Hebrew alphabet to numbers according to this scheme:

Hebrew letter	Rough English equivalent (modern Hebrew)	Numeric value
א (aleph)	glottal stop/no pronunciation	1
ב (bet)	B or V	2
ג (gimel)	G	3
ד (dalet)	D	4
ה (hei)	H	5
ו (vav)	V or O or U	6
ז (zayin)	Z	7
ח (cheit)	Ch (as in 'loch')	8
ט (tet)	T	9
י (yod)	Y	10
ך or כ (kaf)	K or Kh	20
ל (lamed)	L	30
ם or מ (mem)	M	40
ן or נ (nun)	N	50
ס (samekh)	S	60
ע (ayin)	glottal stop/silent	70
ף or פ (pei)	P or F	80
ץ or צ (tsadei)	Ts	90
ק (qof)	Q	100
ר (resh)	R	200
ש (shin)	S or Sh	300
ת (tav)	T	400

The reason for this wide-ranging distribution of numbers is that Hebrew letters were used as numbers, following a system similar to Roman numerals (see p.137), for many centuries, although modern Hebrew, like most languages, uses a Western-style decimal system for pretty much everything. Back then, though, the assignment of letters to numbers had to jump into tens and then hundreds, otherwise such a number system could only go up to 22. Some systems then go on to add numeric values to final Hebrew letters (those at the ends of words, which are sometimes written differently).

Anyway, once the numbers have been assigned, it's time to look at how they fit into words and their meaning. Sometimes this happens in rather an interesting way. Here's an example I stole from the film π (see p.32), though to be fair I did check it:

> The Hebrew word for 'father' is 'ab', spelt אָב.
> (Hebrew is written right-to-left, like Arabic.)
> The gematria value for this, according to the table on page 79 is $1 + 2 = 3$.
>
> The Hebrew word for 'mother' is 'em', spelt אֵם.
> Looking at the table on page 79, we get the gematria value $1 + 40 = 41$.
>
> If we then look at the Hebrew word for 'child' – 'yaelaed', spelt יֶלֶד and take that gematria value we get $10 + 30 + 4 = 44$ – which is the sum of the values for 'mother' and 'father', just as a mother and father are required to create a child.

The writing's on the wall

Another well-known example of gematria is to do with the Hebrew word for live – 'chai', spelt חי with a gematria value of 18. A common expression for charity among devout Jews is 'to give twice chai' – an idea of a charitable life as a double life, living for others as much as yourself. This tradition is often honoured at weddings, bar mitzvahs and other celebrations, where money is traditionally given in multiples of 36 – twice 18 and hence 'twice chai'.

The first person to use gematria was not Jewish (the idea of gematria predates its adaptation in Jewish mysticism, although this is where it's now most popular) but Mesopotamian – Sargon II, ruler of Mesopotamia in the eighth century BC, built the perimeter wall of his palace 16,283 cubits long, to correspond to the gematria value of his name in his language. The ancient Greeks also used a form of gematria (the word is etymologically linked to the word 'geometry') but we don't really have space to look at every kind of gematria.

However, if gematria were really some sort of mystic divination system, it should hold true for pretty much everything, whereas in fact there are hundreds of words with identical numbers that are semantically unrelated. Some people, however, place great faith in this sort of thing, although gematria is generally viewed as an interesting numeric game, rather than something to live by.

Some have made some rather wild predictions with numbers in religion, however. By looking at the letters of the Bible (and the Torah), using numeric sequences to choose the relevant letters from a grid of the original Hebrew, a number of people have claimed to find hidden codes in these holy books which predict the future. Although some have been sort of successful (a period of geopolitical instability from about 2002 onwards, for instance*), the prediction of nuclear apocalypse in 2006 is now significantly overdue. The Number of the Beast (see p.76) may derive from gematria, and a form of gematria is often used to link it back to some perceived evil of the current time.

MODERN NUMEROLOGY

Unlike some of the scientific ideas mentioned elsewhere, numerology occupies the most irrational end of the spectrum of ideas and theories about numbers. The practice of numerology has been going on in some form for many thousands of years – the system of lucky and unlucky numbers still used in China, for instance, (see p.88) goes back many centuries. Modern numerology in the West, however, began in the late twentieth century, drawing on various historical traditions, including

* On the other hand, when has the world ever really been stable?

that of the Greek mathematician Pythagoras and Kabbalistic gematria (see p.79).

Modern numerology assigns very vague significance to various digits, usually only the digits 0–9. The meanings shown here are common ones from several different schools of thought, and hence may appear self-contradictory as well as sometimes rather obvious:

0 – everything, existence, beyondness, universality

1 – solitude, unity, independence, initiative

2 – duality, division, cooperation

3 – movement, synthesis, divinity
(in Christian tradition – see p.103)

4 – solidity, resistance, materiality

5 – life, love, growth, regeneration

6 – union, perfection, wholeness, imperfection of man
(in Christian tradition – see p.78)

7 – magic, mysticism, wisdom, divinity
(see Seven, p.70)

8 – good luck (especially in China), practicality,
justice, power

9 – major change, finality, achievement

There's also a numerological theory revolving around the number 23 (see p.60).

It all adds up

The actual practice of numerology generally involves analysing numbers as they appear in the subject's life, particularly where they recur. A form of gematria (see p.79) is particularly popular, often involving adding up the letters in people's names. In this example, I'm using a popular system where the letters are numbered only up to 9, after which the numbering repeats again (so 1 could be A, or J, or S, etc.), but there are as many methods of assigning numbers to letters as there are interpretations of the numbers:

J	A	M	I	E		B	U	C	H	A	N
1	1	4	9	5		2	3	3	8	1	5

The numbers are added together in a process known as digit summing. This can be done as many times as necessary to produce a single digit, since single digits are used frequently in everyday life and hence provide more convenient coincidences. I mean, single digits are more magical.

1 + 1 + 4 + 9 + 5 + 2 + 3 + 3 + 8 + 1 + 5 = 42 (I quite like the *Hitch-Hiker's Guide to the Galaxy* associations, anyway)

4 + 2 = 6

So, rather confusingly, I'm either perfect or deeply imperfect. Or related in some way to notions of union. Or something.

Ultimately, numerology is concerned with nothing more than finding absurdly vague meaning in numeric coincidence. The faculty of pattern recognition is often too easily fooled by coincidence.

Numbers rule the universe

Numerologists sometimes cite Pythagoras' axiom that numbers 'rule the universe' as the origin of their discipline, but this seems almost a perversion of mathematical values. Pythagoras was a mathematician as well as a mystic, and is certainly more famous today for the Pythagorean Theorem* than his mysticism. To me, the vague meanings and coincidences of numerology seem to be no match for the fascination of rational mathematical concepts like the Fibonacci sequence, which demonstrate in scientific terms how numbers really rule the universe.

One might argue that humans have believed in some form of nonsense for generations, and if they want to believe it, where's the harm? But when people believe such irrational

* The Pythagorean Theorem states that, in a right-angled triangle, the square of the longest length (the hypotenuse) is equal to the sum of the squares of the other two lengths.

things there's always the danger that they'll act on them, and such was the case in a murder trial in Washington DC in 2006. According to the *Washington Post* report of the story, the trial had already taken two strange turns – a prosecution witness testifying while under the influence of cannabis and, more sinister, a juror being tracked down by a major figure in the case – when the jury reported that one of their number was using numerology rather than evidence to determine her position. The juror was eventually removed, and two men were found guilty of the murder.

Bent as a Nine-kyat Note

An even more bizarre story of faith in numerology comes from 1980s Myanmar (Burma). The country has been ruled by a military junta since 1962, and between 1974 and 1988 was effectively ruled (but not ruled effectively) by General Ne Win, who suppressed political dissent with autocratic laws and military force. Ne Win's economic policy was especially disastrous, characterized by extreme isolationism. Not only that, but he was also corrupt – quite literally as bent as a nine-bob note.

The general was a keen numerology enthusiast, and when his personal astrologer and numerologist advised him that his lucky number was 9, and that he'd live to be 90 if he placed the number all around him, he announced one day that he was reissuing the kyat (Burma's currency) in notes which were multiples of 9, and that, by the way, everyone's decimal-system life savings were now entirely worthless. In a rather pleasing numeric twist, this conversion was one of the major catalysts for the '8888 Uprising', which began on 8 August 1988. Ne Win, however, got his wish – he died in 2002, at the age of 91.

CHINESE LUCKY (AND UNLUCKY) NUMBERS

The nature of Chinese languages, with many very similar-sounding words, leads to most digits being considered lucky or unlucky, usually because of homophony (words sounding similar to other words when spoken) in the same way as text speak uses 2 for 'to', 4 for 'for', etc. This is a non-exhaustive list of them (which vary significantly across China's 1.3 billion inhabitants and numerous dialects).

1

Much as in the West, the number 1 can be interpreted as representing either unity or loneliness.

2

The number 2 sounds like the word 'easy' in the Cantonese language spoken in much of China, but the idea of a pair has wide-ranging importance. Doubling – 'double happiness' or 'double prosperity' – is a common way of intensifying good wishes.

3

Homophonous with the word 'life', the number 3 is a rather auspicious number, especially when combined with other lucky numbers.

4

An unlucky number across the Far East, in much the same way as 13 (see p.100) is in Western cultures. Four, and by extension all numbers containing the digit, is associated with death because the two words sound similar in most dialects. Many Chinese-made products, including Nokia phones and Canon cameras, skip from 3 to 5 in model numbers. In some areas, however, it sounds closer to the word 'task' and is considered lucky for this reason.

5

The number 5 is associated with the five elements (fire, earth, air, water and the aether), and also, in Mandarin, with the word meaning 'I' or 'me'. In Cantonese, however, 5 sounds closer to the word 'not'.

6

While 666 is still the subject of Satanic superstition in the West (see Number of the Beast, p.76), it's routinely displayed in Chinese shop windows. The number 6 sounds like the word for 'smooth' or 'easy', and repeating it three times maximizes the effect.

7

Although the number 7 is associated with death and tragedy, it does not carry the same dread as 4. It also has associations with togetherness, expressed in the folk tale of the Cowherd and the Weaver Girl, two lovers forced apart by the gods and permitted to meet only once a year, on the seventh day of the seventh lunar month. In China, this day is celebrated as *Qi Xi* – similar to Valentine's Day, with an element of supernatural tragedy – and similar celebrations occur elsewhere in Eastern Asia.

The seventh month of the Chinese calendar (generally around August) is 'The Month of the Hungry Ghosts', when the spirits of the dead wander among the living. It isn't a wholly inauspicious time – many families pay homage to dead ancestors, but it's considered wise to try not to annoy them. Many people leave gifts for the ghosts and avoid disturbing them by travelling or moving house.

8

The start time of the Beijing Olympics – around 8:08pm on 8 August 2008 – reflects China's relationship with the number 8. It's something of a national obsession – car number plates and telephone numbers containing eights can command thousands of dollars. The root of this phenomenon is partly in homophony, with the word 'eight' sounding similar to 'wealth' or 'prosper', but there's also a deeper philosophical meaning. The number 8 resonates with the Buddhist Eightfold Path, that defines eight morally correct courses of action for Buddhists, who make up about half of the Chinese population.

The other major Chinese philosophies, Confucianism and Taoism, also have eight at their root. The Bagua is a fundamental philosophical concept that breaks the universe into eight elements whose numerous manifestations include weather, natural formations, family members, compass points and moods. As such, eight is not simply a number of wealth, but one that touches the very root of China's spiritual life.

9

Like 6 and 8, 9 is considered lucky. Through homophony, it's associated with stability and lasting happiness, and hence is much in evidence at weddings and similar major life events. In Cantonese, it's

> roughly homophonous with 'sufficient', which is similarly auspicious. As the highest single-digit number, it's also associated with greatness and particularly the emperors of China.

These numbers are especially lucky (or unlucky) when combined with each other. Repeated lucky digits are preferable to single ones, as this intensifies their effect, and combining numbers can create especially lucky phrases. For instance, 168 can be taken to mean 'smoothly prosperous together' – a good omen for any business deal – while even 4 can be redeemed in this way – 54 in Cantonese suggests immortality ('no death') – but is especially unlucky when combined with 1 into 14, suggesting one's own death, or dying alone. Many high-rise buildings skip the fourteenth floor (and sometimes all floor numbers with 4 in them), as well as increasingly adopting the Western custom of skipping the thirteenth. There are marketing advantages here as well, of course – a shrewd Chinese developer can sell a fiftieth-floor flat in a thirty-six-storey building.

ZODIACS

Western zodiac

The traditional Western zodiac consists of twelve symbols, each corresponding to a period of roughly a month, based on the position of the sun relative to certain constellations in the sky, which changes across the year. People born at certain times are assigned 'star signs' based on their date of birth, which are associated with vaguely defined personality traits:

Aries (Ram) – 21 March to 20 April – Energetic, enthusiastic and confident, but may also be quick to anger, impatient and impulsive.

Taurus (Bull) – 21 April to 21 May – Reliable, placid and loyal but possibly possessive and stubborn.

Gemini (Twins (Castor and Pollux)) – 22 May to 21 June – Versatile, eloquent and witty but, on the other hand, may be nervous and scheming.

Cancer (Crab) – 22 June to 23 July – Loving, protective and intuitive, but prone to be over-emotional and clingy.

Leo (Lion) – 24 July to 23 August – Enthusiastic, generous and creative, but also associated with pompousness, interference and bossiness.

Virgo (Maiden) – 24 August to 23 September – Modest, reliable and intelligent, but with a fussy, conservative side.

Libra (Scales) – 24 September to 23 October – Diplomatic, sociable and idealistic, but sometimes indecisive and easily influenced.

Scorpio (Scorpion) – 24 October to 22 November – Determined and passionate, but might be jealous, obsessive and deceitful.

Sagittarius (Archer) – 23 November to 21 December – Optimistic, honest and philosophical, with the potential for carelessness and irresponsibility.

Capricorn (Goat or ibex) – 22 December to 20 January – Prudent, careful and reserved, but perhaps also pessimistic and grudging.

Aquarius (Water bearer) – 21 January to 19 February – Friendly, intelligent, loyal and inventive, but contrary and prone to detachment.

Pisces (Two fish) – 20 February to 20 March – Sensitive, unworldly and sympathetic, but might be escapist, vague and weak-willed.

There's also a thirteenth sign – 'Ophiuchus' the serpent-bearer – which is not recognized by most astrologers, perhaps to avoid the unlucky association of 13 signs, or to keep it down to a conveniently composite 12.

The zodiac is subdivided in a number of ways to make use of 12's divisibility – into two genders (rather oddly, Aries, the ram, is masculine while Taurus, the bull, is feminine), three 'qualities' (cardinal signs at the start of a season, mutable at the end and fixed in the middle of a season) and according to the four classical elements (supposedly, people of certain element signs are especially compatible with those of the same element). As unscientific as astrology is, it was a genuine and respectable discipline in less enlightened times – the movement of constellations across the sky seemed a confirmation of Earth's place not at the centre of the universe, but as part of a wider celestial system. It was only after the Renaissance that a distinction between superstition and the science of astronomy began to be drawn.

Signs	**Elements**	**Qualities**	**Gender**
Aries	Fire	Cardinal	Male
Taurus	Earth	Fixed	Female
Gemini	Air	Mutable	Male
Cancer	Water	Cardinal	Female
Leo	Fire	Fixed	Male
Virgo	Earth	Mutable	Female
Libra	Air	Cardinal	Male
Scorpio	Water	Fixed	Female
Sagittarius	Fire	Mutable	Male
Capricorn	Earth	Cardinal	Female
Aquarius	Air	Fixed	Male
Pisces	Water	Mutable	Female

Supposedly, your star sign says a great deal about your personality, but the scientifically verifiable link between the movements of planets and stars and human affairs on Earth is non-existent. It hardly needs to be said again, but much like numerology (see p.82), astrology is manifestly nonsense, which relies on harnessing credulity in vague predictions (invariably positive) for financial gain. Regrettably, this doesn't stop it being a major industry that makes millions for its practitioners.

The Chinese zodiac

Equally unscientific, but perhaps more interesting by virtue of its novelty, is the Chinese zodiac, which assigns one of 12 animals to your birth year (although it does, of course, go by the Chinese year, which usually begins in late January or early February). Which animal are you?

Rat	1900, 1912, 1924, 1936, 1948, 1960, 1972, 1984, 1996
Ox	1901, 1913, 1925, 1937, 1949, 1961, 1973, 1985, 1997
Tiger	1902, 1914, 1926, 1938, 1950, 1962, 1974, 1986, 1998
Rabbit	1903, 1915, 1927, 1939, 1951, 1963, 1975, 1987, 1999
Dragon	1904, 1916, 1928, 1940, 1952, 1964, 1976, 1988, 2000
Snake	1905, 1917, 1929, 1941, 1953, 1965, 1977, 1989, 2001
Horse	1906, 1918, 1930, 1942, 1954, 1966, 1978, 1990, 2002
Ram	1907, 1919, 1931, 1943, 1955, 1967, 1979, 1991, 2003
Monkey	1908, 1920, 1932, 1944, 1956, 1968, 1980, 1992, 2004
Rooster	1909, 1921, 1933, 1945, 1957, 1969, 1981, 1993, 2005
Dog	1910, 1922, 1934, 1946, 1958, 1970, 1982, 1994, 2006
Pig	1911, 1923, 1935, 1947, 1959, 1971, 1983, 1995, 2007

TWELVE

Perhaps the best-known example of the religious significance of the number 12 is the Twelve Apostles of Jesus. (Judas Iscariot's presence as the thirteenth person at the Last Supper may also be the origin of the superstition surrounding the number 13 (see p.100).) The holy books of Judaism and Christianity refer to the Twelve Tribes of Israel, descended from the twelve sons of Jacob. This may well have influenced the Twelve Days of Christmas, the period lasting from Christmas to the eve of Epiphany (Twelfth Night).

Islam was split into two major branches – Sunni and Shia – by a conflict over the rightful succession to the Prophet Muhammad. The vast majority of Shi'ite Muslims believe the succession runs through twelve Imams, the last of whom was born in AD 868 and is, impressively, still alive, but hidden to the world. It should be stressed that the Imams are not prophets, but simply carry out the Prophet's message. This subgroup of Shi'ites are known colloquially as Twelvers, and make up most of the population of Iran and Iraq, as well as being numerous in most other Muslim countries.

Part of this ubiquity must have resulted from 12 being a

highly composite number – one with many factors that can be divided up in many different ways. Twelve is the smallest number with six factors – 1, 2, 3, 4, 6 and 12. The Duodecimal Society (now known as the Dozenal Society), founded in 1944 in New York, tried to promote the adoption of a duodecimal (base-12) number system to take advantage of this quality. Unsurprisingly, they were entirely unsuccessful.

The number of factors in the number 12 has made it convenient for measuring time for thousands of years. The calendar has 12 months, although the actual number of lunar cycles in a year is closer to 13. The lunar cycle's rather too inconvenient to use for time measurement, though, so 12 months are used to fit it in with solar years. Much of this derives from the Babylonian time system, which used 360 days of the year (which is also the origin of our circle being divided into 360 degrees). Twelve months are conveniently divisible into four seasons of roughly three months each. Each day is then divided into 24 hours (a practice that originated with the Sumerians splitting day and night into 12 parts each), and each hour and minute are divided into 60 parts – a multiple of 12.

Western and Eastern zodiacs (see p.93) make use of 12 signs, which again tend to be divided into categories to make use of 12 being composite. Twelve is also traditionally the number of people on a jury, although there seems to be no particular reason for choosing that number other than convenience, and possibly because 12, being even, allows for a jury opinion to be split exactly in half, although modern jury trials generally require most of the jury to agree on a verdict.

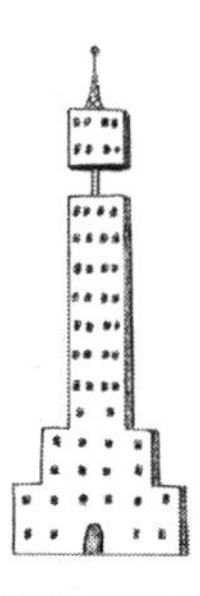

THIRTEEN

From Biblical times to modern air travel, 13 is the West's unluckiest number. This superstition supposedly derives from the number present at the Last Supper – a Biblically resonant 12, plus a treacherous thirteenth – Judas Iscariot – who was supposedly the last to sit. A similar Norse myth involves 12 gods sitting at a table in Valhalla, only for the treacherous Loki to turn up and kill one – Baldur, the heroic god of beauty.

Thirteen may also be unlucky in Christian traditions simply because it is holy to some pagan religions, whose calendars contain 13 lunar months. At a more abstract level, it may be that the real reason has to do with 13's mathematical awkwardness – while 12 is recognized in many cultures for having many factors, 13 right after it is awkwardly prime (see p.122). It seems likely that more than one of these theories is at least partly correct, given that the superstition about 13 doesn't really make an appearance until the Middle Ages, well after the ill-fated meals of Biblical and Norse mythology.

The fear that Friday 13th, specifically, will bring bad luck, seems to be more recent, but linked to a long tradition of superstition surrounding Fridays generally, including Good Friday and the British tradition of public hangings on Fridays. Some historians suggest that the day and the number were only brought together in 1907, by Thomas W. Lawson's little-known novel *Friday, the Thirteenth* (which is apparently about turn-of-the-century Wall Street and not a hockey-masked serial killer). Another theory, recently lent popularity by *The Da Vinci Code*, is that it derives from Friday 13 October, 1307, when King Philip IV issued a death warrant for the Knights Templar.

Although there's obviously no verified causal link between Friday 13th and bad luck, the date might lead indirectly to cause for concern. A number of studies have found higher rates of accidents, including car crashes, on Friday 13th. These could, however, be due to Friday 13th nerves distracting drivers, as well as a degree of confirmation bias – those who believe Friday 13th is unlucky are more likely to attribute even slight bad luck to the date.

Despite the comparative lack of superstition in modern society, triskaidekaphobia (fear of the number 13*) remains prevalent. At dinner parties and wedding receptions the hosts will go to great lengths to avoid seating 13 guests at a single table, as, so the superstition goes, one of the guests will die within a year. Tenants in many skyscraper office blocks will notice that the floors skip from the 12th to the 14th (or floor 13 has been called 12A), and airlines frequently do the same with their seating rows. In China, tetraphobia (fear of the

* The fear of Friday 13th specifically is known as paraskavedekatriaphobia.

number four) is even more pervasive – see p.88 for more on China's lucky and unlucky numbers.

Sensing the absurdity of this superstitious fear in a supposedly rational society, Captain William Fowler, an American Civil War veteran, set up the Thirteen Club in New York in 1880. Captain Fowler's life featured an improbable recurrence of the number 13 (including his early education at Public School No.13, his construction of 13 public buildings as an architect and his experience of 13 Civil War battles), and he came to view it as lucky rather than cursed.

The highly exclusive Thirteen Club's dinner parties always had 13 people at each table, and its members included future President Theodore Roosevelt and former incumbent Chester Arthur. A later imitator, the London Thirteen Club, was similarly well appointed, counting a number of MPs among its members. Other superstitions were also ridiculed – members might enter the room by passing under a ladder, before deliberately spilling salt during dinner and breaking mirrors afterwards.

THREE

In the West, the Holy Trinity of the Father, the Son and the Holy Ghost is probably the best example of a number possessing religious significance. The superstitious fear of walking under a ladder seems to have its origin here – a ladder propped against a wall forms the longest side of a triangle with the ground and the wall forming the other two sides, and a person passing under the ladder is symbolically breaking the Trinity and bringing themselves bad luck. The Holy Trinity is by no means unique in assigning significance to the number three, though. The Bible describes Jesus receiving three gifts from three wise men, preaching for three years and rising three days after his death. Celestial realms are often divided into three levels – Heaven, Hell and Purgatory.

Islam, the third of the three Abrahamic religions, has three holy cities – Mecca, Medina and Jerusalem. The fairly numerous gods of the Hindu religion are headed by a trinity (*trimurti*) of Brahma (creator of the universe), Shiva (destroyer and recreator) and Vishnu (wise protector). Ancient Greek mythology featured three Furies, three Muses (which later increased to nine) and three Gorgons (most famously the snake-haired Medusa) among other trinities. In the neo-pagan Wiccan religion, the Rule of Three dictates that a person's actions, good or bad, will be visited threefold upon them, much like the concept of karma in Hinduism and Buddhism.

The mystic significance of the number three may be linked to ideas of completeness and harmony – if two suggests a duality in conflict, three might suggest a more stable balance, particularly since the triangle is the simplest, and thus the most structurally efficient and geometrically common two-dimensional shape.

Public speakers and mottoes frequently list three things in a technique known as tricolon – 'life, liberty and the pursuit of happiness', 'liberty, equality, fraternity', 'education, education, education'*. In the visual arts, a triptych is an arrangement of three panels, traditionally part of an altarpiece but also adopted by secular artists for its striking effect. Francis Bacon's *Three Studies for Figures at the Base of a Crucifixion*, painted in 1944, is an excellent and very influential example.

Three is a good number for stories as well, where the third

* A tricolon of tricolons.

person can act as a balancing element between two conflicting parties, or as someone who upsets the balance, such as in a 'love triangle' scenario. Traditionally, stories are said to consist of three parts – an introductory beginning, a middle that further develops the story, and an end that concludes it, and trilogies of books and films often work in much the same way. In children's stories, the number three (little pigs, bears, etc.) allows the beginnings of a pattern to be set with the first two before the pattern is overturned with the third (the house is still standing, Goldilocks is discovered). Jokes, particularly those at the expense of certain nationalities or hair colours, often use a similar format.

FOUR

Although not really possessed of the same mystique as three (see p. 103), four nonetheless holds an interesting place in the world's culture of numbers, probably at least partly because it's an even number and the smallest perfect square (see p. 123) above one, and also perhaps because humans have four limbs and four fingers (not counting the thumb) on each hand.

Four humours

The number four appeared in Ancient Greece as the four classical elements – earth, water, air and fire. Similar systems of several different elements were used in other cultures, such as Hinduism and Chinese Taoism, which had five – fire, wood, water, metal and earth.

The classical elements were believed to form all earthly matter when combined together, but there was also a fifth special element, generally known as the 'idea' or the 'aether'. All heavenly bodies were made from aether, and it was the medium in which the gods lived. The word 'quintessence', meaning a completely pure quality or archetype, is derived from the fifth element in this theory. The idea of the four (or five) elements continued to influence science for centuries, lasting through the Middle Ages.

Even after the classical elements had been dropped, scientists continued to believe up until the twentieth century that the universe was filled with a substance called aether. Although we now know that the universe is mostly empty and that light can travel through a vacuum, these discoveries came as an enormous surprise to the scientists who made them, who assumed that light, like sound, could only travel through some sort of medium.

The four elements mapped neatly onto the four 'humours' of ancient medicine, a theory based on bodily fluids. Ancient Greek and Roman doctors reasoned that having too much of one of four fluids (blood, phlegm, yellow bile and black bile) would cause ill health, and associated each imbalance with a personality type or 'temperament'. Despite the complete refutation of this theory, it continues to influence English adjectives for personalities.

Blood (associated with air) – sanguine personalities are considered outgoing, friendly and confident, but with the potential to become arrogant and impulsive, finding it difficult to concentrate on one thing.

Phlegm (water) – phlegmatic people are calm and rational, and able reliably to keep a cool head, but often dispassionate and unenthusiastic to the point of laziness.

Yellow bile (fire) – 'choleric' tends to denote practical-minded, passionate, dynamic types whose greatest fault is generally their suspicion and quickness to anger. Cholera, a dangerous disease contracted from infected water supplies, was once thought to be caused by bile.

Black bile (earth) – melancholic personalities are, obviously, those overly preoccupied by sadness and tragedy. Melancholic people were considered the most creative and poetically minded, though. Basically, goths.

This theory continued to affect medicine right up to the

nineteenth century, and is the origin of bloodletting – bleeding a patient who has been diagnosed as excessively sanguine. Although the four humours theory of medicine is long dead to all but the craziest quacks, a similar concept of four temperaments has been applied in modern psychoanalysis. Carl Jung (1875–1961), a pioneering psychoanalyst who broke away from Freudian tradition to embrace some occasionally rather odd ideas, viewed the idea of 'fourfoldness' as one of humanity's most important and profound cultural building blocks. Jung's views of people as being divided into introverts and extroverts influenced the development of various personality classifications over the following decades. In some ways, the four temperaments are very much alive.

Four horsemen

The Four Horsemen of the Apocalypse are described in the Book of Revelation, the last book of the Bible – which also gave us the Number of the Beast (see p.76) – as the harbingers of the end of the world. Only Death, riding a pale horse, is named in the text, but from the description the others are generally considered to include War, on a red horse and carrying a sword, Conquest, a regal figure on a white horse, and another horseman bearing scales and measuring out food, who is generally considered to represent Famine. In many interpretations, Pestilence replaces Conquest, who has a rather large overlap with War anyway.

More recently, the term 'Four Horsemen' has been used, usually humorously, to refer to the four leading authors of the 'New Atheist' movement: Richard Dawkins, Daniel Dennett, Sam Harris and the late Christopher Hitchens. It hasn't yet been suggested which author is meant to be which horseman.

Other instances of the number four in religion are the four official Gospels (the books of Matthew, Mark, Luke and John, which kick off the New Testament with an account of Jesus' life) and the four rivers that flowed through Eden in the Old Testament (the Pishon, the Gihon, the Tigris and the Euphrates).

FIVE

As with several other numbers already mentioned, the number five is important to various religious faiths. In Islam, there are five sacred duties for all Muslims, known as the Five Pillars of Islam:

Shahaadah –	a declaration of faith in God alone, and in Muhammad as his prophet
Salah –	prayers five times daily
Zakat –	giving to charity
Sawm –	fasting from dawn to dusk during Ramadan
Hajj –	a pilgrimage to Mecca, which should be made at least once in a Muslim's lifetime

Some branches of Islam prescribe more than five duties, adding a sixth or seventh pillar.

Five is also important in Judaism, where the Torah is composed of five books – the 'Pentateuch' of Genesis, Exodus, Leviticus, Numbers and Deuteronomy. In Sikhism, Sikhs have to wear the symbolic 'five Ks' - *kesh* (long hair), *kara* (steel bracelet), *kanga* (comb), *kachha* (cotton breeches) and *kirpan* (a ceremonial sword or dagger*). Sikhism has five Virtues and five Evils, similar to the Seven Deadly Sins and Heavenly Virtues (see p.70):

Virtues	**Evils**
Sat – truth	*Kaam* – lust
Santokh – contentment	*Krodh* – anger and hatred
Daya – compassion	*Lobh* – greed and covetousness
Nimrata – humility	*Moh* – attachment to worldly things
Pyare – love (including love of God)	*Ahankar* – pride and egotism

* There have been a number of legal controversies in the UK and abroad where a Sikh's requirement and religious right to wear the *kirpan* has conflicted with the law on carrying weapons. Although the *kirpan* is always worn in a sheath, and intended as a symbol to demonstrate a commitment to non-aggressive behaviour (traditionally, the *kirpan* was a defensive weapon), one can understand the concern of the authorities. Various solutions have been adopted, including *kirpans* with very short blades, blunting the blade and bolting the *kirpan* into its sheath.

The pentagram

No discussion of five's religious significance would be complete without a look at one of the world's oldest, most controversial and most versatile symbols – the pentagram.

At least 6,000 years old, the pentagram may have originated in ancient Sumer as a pictogram for the heavens, and subsequently been adopted by the Babylonians. The Pythagoreans (followers of the Greek mathematician, philosopher and mystic Pythagoras of Samos (around 580–500 BC)) viewed the pentagram with a certain reverence: not only could the five points be considered representative of the five classical elements (earth, fire, water, air and the aether), but the main lines of a geometrically regular pentagram are divided in the Golden Ratio (see p.44).

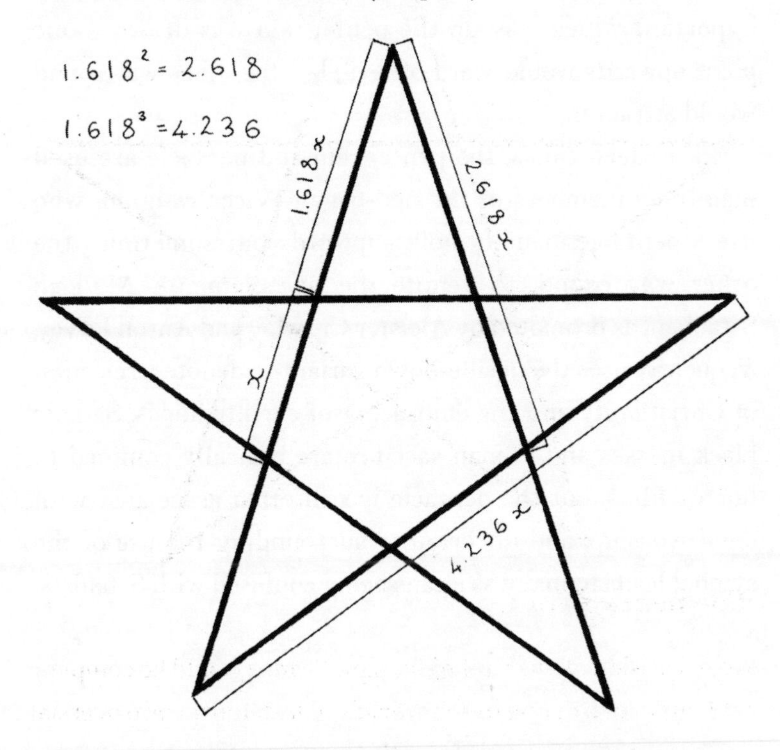

Despite the common occult and even Satanic associations of the pentagram (which is known as a pentacle when it has a circle round it), it was a popular Christian symbol for many centuries. It symbolized the Five Wounds of Christ (one in each hand, one in each foot, and one in the side) and was often daubed on doors to protect against witches, demons and other evil spirits.

In this way, the pentagram became associated with the occult, often linking back to the Pythagorean view of it as representative of the elements. Eventually it became very important which way up the pentagram was drawn – one point upwards would ward off evil, but the other way round would attract it.

In modern times, the pentagram and pentacle are used mainly by members of the neo-pagan Wicca religion, who use a pentacle, usually point-upwards but sometimes the other way round, to denote the five elements. Modern Satanism, as promoted by Aleister Crowley and Anton LaVey, frequently uses the upside-down variant to denote a rejection of Christianity and the embracing of earthly ideals. Satanic black masses and human sacrifice are basically confined to horror films, but the pentacle is still often associated with devil-worship, and uncertainty surrounding the use of the symbol leads to many Wiccans being confused with Satanists.

NUMBERS IN MATHS AND SCIENCE

A MATHEMATICAL GLOSSARY

Before we go deeper into the maths, it might be an idea to review some of the mathematical terms you've probably learned at school ...

Average

A number – actually several different numbers – that can help to summarize the distribution of statistical data in an easily readable way. There are three main types of average, and numerous more complex ones.

The mean – generally what's meant by 'average'. It's a 'middle' value, which is calculated by adding together all the values and then dividing that by the number of values.

The mode – simply the most common value in any set of data. The modal number of legs per person in the UK is two, because most people have two legs. (The mean is a little below two, as it accounts for people with fewer than two legs).

The median – the middle value in a set of data: the point that separates the lower half from the higher half. It's calculated by ranking the data in ascending or descending order and then choosing the middle value, or, where there's an even number of values (and hence no middle one) the mean of the two middle values.

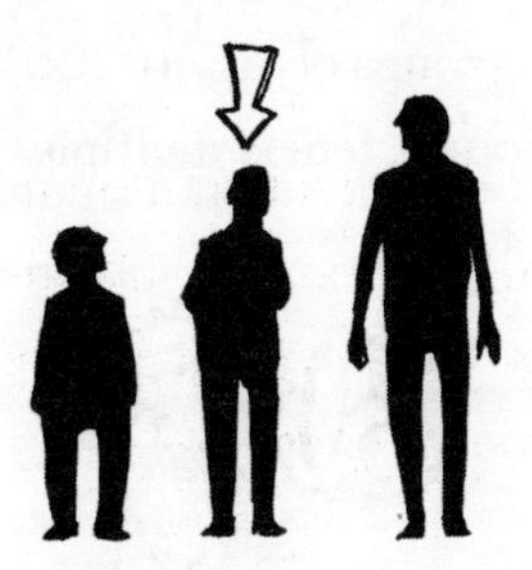

Base

In number systems that use positional notation (like ours), the base is the number defining the relationship between the positions. 'Positional notation' refers here to the fact that our numbers make use of a small number of symbols (the digits 0–9) but these are then interpreted according to their position in a number, i.e. we know 900 is bigger than 9 because the digit 9 occupies the hundreds column in 900 but the ones column in 9.

Although this system has been drummed into us all since primary school, positional notation wasn't always the norm. The best-known example of non-positional numbering is the Roman numeral system (see p.137). Anyway, the 'base' is the number that links the 'positions' of positional notation – in a base X system, a number written 10 is X times bigger than 1, and each step to the left moves one up the powers (see below) of X.

Our decimal system uses base 10 ('decimal' comes from the Latin for 'ten'), so children learn to think of complex numbers as 'hundreds, tens and units'. A more grown-up approach would be to see it as groups of 10^2, 10^1, 10^0 (the last column in any positional notation system is multiples of one, as anything to the power of 0 is 1).

A number of other positional systems have been used in the past: the Babylonians used a sexagesimal one (base 60*), which isn't to be confused with hexadecimal – a base-16 number system used by some computer applications, with the digits 0–9 being extended with the letters A–F. Because 16 is a power of 2, the hexadecimal system can be used easily to represent numbers from the base-2 binary system (see p.150).

* In our decimal (base-10) system, the digits are related by powers of ten. Similarly, a base-60 system is based around powers of 60, with the rightmost digit representing multiples of one (any number to the power of zero is one), the next multiples of 60 (60^1), the next multiples of 3600 (60^2), and so on. Such a large base is complex enough by itself, but added to this was the Babylonian method of writing the numbers 0–59, which was sort of decimal – a single symbol signifying 10 could be repeated and combined with other digits in a similar way to Roman numerals, within the sexagesimal system.

Incidentally, the word 'hexadecimal', like 'television', has a mix of Greek and Latin roots – so, why not call it sexadecimal, and be consistent with 'sexagesimal'? From what I understand, the IBM managers of the 1950s considered 'sexadecimal' too rude!

Cube number (or perfect cube)

A number equal to the cube of an integer (see p.121) – the integer times itself twice (for example: $3 \times 3 \times 3 = 3^3 = 27$, the third cube).

The first 10 cubes are:

1, 8, 27, 64, 125, 216, 343, 512, 729, 1000

Degree

The most common way of measuring the angle between two lines, represented by this symbol: °. There are 360 degrees in a circle, which might seem a rather odd number to use. The reason for this is partly that 360 is a highly composite number (with 24 factors that divide into it), but also because the ancient Babylonians who divided the circle into degrees (though the 360° circle appears to have been popularized by the Greeks) used a 360-day year and a number system with base 60 (60 being one-sixth of 360). For the most precise measurements of angle (such as measuring exact latitude and longitude, or the accuracy of a rifle), the degree is subdivided into 60 minutes of arc, which are each made up of 60 seconds of arc.

The number of factors in 360 allows the circle to be easily divided into various parts, such as the right angle of 90°, which is exactly a quarter of the full circle, or 60°, which is the angle found in equilateral triangles. By contrast, the radian, the second most common measurement of angle, is nothing like as intuitive, being defined as $1/(2\pi)$ of a circle, or around 57.3°. This is all very well for calculus but a bit awkward for ordinary geometry.

Factor

A number that divides into another number. Some numbers have many factors, and hence are said to be highly composite, while prime numbers have none but 1 and themselves. Twelve, for instance, is a highly composite number, while 13, a prime, is not.

Factors of 12: 1, 2, 3, 4, 6, 12
Factors of 13: 1, 13

Integer

This is simply another word for a whole number, i.e. one that doesn't have to be expressed as a fraction.

Irrational number

A number which, unlike rational numbers, cannot be expressed as a fraction. Irrational numbers are non-repeating decimals with infinite decimal places. Square roots of integers that aren't perfect squares, such as:

$$\sqrt{47} = 6.8556546004010441249358714490848 \ldots$$

are irrational. However, this number could be approximated to a rational one if we rounded it to a finite number of decimal places.

$$6.85565 = 6\ {}^{85565}/_{100000}$$

Prominent examples of irrational numbers include φ (see p.46) and π (see p.129).

Natural number

A positive whole number of the type that occurs in nature. You can have one stone or two stones or three, but not half a stone or -7 stones. You can also have 0 stones, of course, but it remains undecided in the mathematical world whether 0 can be considered a natural number.

Powers

A number multiplied by itself a certain number of times, that second number being known as the power or the exponent, and written in superscript, $^{\text{like this}}$. Squared and cubed numbers are respectively second and third powers of the original number. Positional number systems are based around powers – in our decimal system the digit positions refer to powers of 10 – thousands, hundreds, tens etc. Any number to the power of 0 equals 1, and of course any number to the power of 1 is the number itself.

Prime number

A number that is divisible only by itself and 1 (1 is not a prime number, however). Prime numbers are useful in cryptography, because large prime numbers can be multiplied together into much larger numbers, and it's then very complicated to work out what the original prime numbers were. Because of this application of prime numbers, working out patterns in their distribution remains a major mathematical project. The first ten prime numbers are:

2, 3, 5, 7, 11, 13, 17, 19, 23, 29

Rational number

A number that can be expressed as a fraction, with integers on the top and bottom. If written as a decimal, a rational number will either terminate (i.e. will have a finite number of decimal places) or repeat (the digits after the decimal point will follow an obvious sequence). For example, 1 (1.0), ½ (0.5), ⅓ (0.33333333333...) and $\frac{355}{389}$ (0.9125964010282776) are all rational numbers.

Square number (or perfect square)

A number equal to an integer squared (i.e. multiplied by itself). 2 times itself is 4, so 4 is the second square. The square of a negative number is equal to the square of its positive counterpart, i.e. $-3 \times -3 = 3 \times 3 = 9$. The first ten square numbers are:

1, 4, 9, 16, 25, 36, 49, 64, 81, 100

Square root

The inverse of squaring a number, the square root of a number *a* is the number which, when multiplied by itself, will produce *a*, and is denoted by the symbol $\sqrt{}$. $\sqrt{25} = 5$ because $5^2 = 25$. In fact, though, any positive number has both a positive and a negative square root, as $(-5)^2$ is also 25. Negative numbers, however, have no real square roots although they do have imaginary ones (see i, p.142).

Triangular number

One of a set of integers which correspond to the number of points required to produce a triangle, where each row of points has one more than the previous row. A triangular number will be the sum of a series of consecutive integers, starting from 1. Thus, the nth triangular number will equal n+(n-1)+(n-2)+(n-3) ... until you get to 1.

The first ten triangular numbers are:

1	= 1
3	= 1+2
6	= 1+2+3
10	= 1+2+3+4
15	= 1+2+3+4+5
21	= 1+2+3+4+5+6
28	= 1+2+3+4+5+6+7
36	= 1+2+3+4+5+6+7+8
45	= 1+2+3+4+5+6+7+8+9
55	= 1+2+3+4+5+6+7+8+9+10

Or, more graphically:

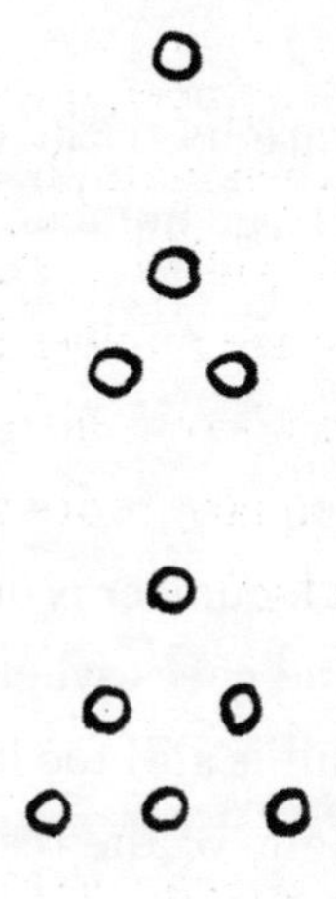

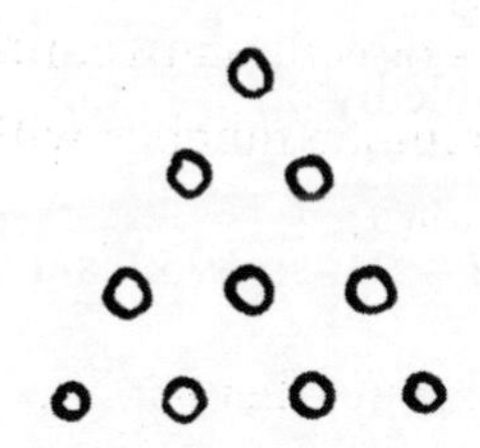

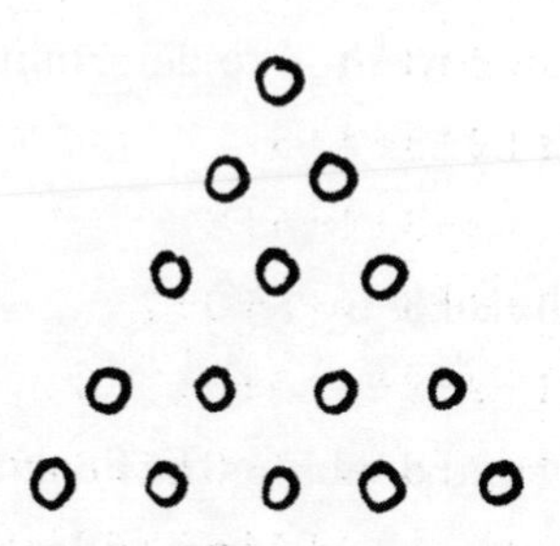

DIVISIBILITY TRICKS

In mental arithmetic, various tricks can be used to quickly check the divisibility of large numbers by smaller ones. Here are a few examples.

Divisibility by 3

To check whether or not a number is divisible by 3, simply add all the digits together and check whether the sum of these is divisible by 3. If the result is still too large to be conveniently divided mentally, the digits of the result can be added again until it's clear (if the number is divisible, you should be able to keep going until you end up with 3, 6 or 9):

Is 5952 divisible by 3?

$5 + 9 + 5 + 2 = 21$, so 5952 is divisible by 3.

Divisibility by 7

To be honest, it's usually quickest to work this out with long division, but there is a method to determine whether or not a number is divisible by 7.

Is 21987 divisible by 7?

Begin by removing and doubling the final digit of the number.

$7 \times 2 = 14$

Subtract this from the remaining digits:

$2198 - 14 = 2184$

Do the same again:

$4 \times 2 = 8$

$218 - 8 = 210$

And keep going until you get to a multiple of 7:

$0 \times 2 = 0$

$21 - 0 = 21 = 3 \times 7$, so 21987 is divisible by 7.

Divisibility by 9

The same as the rule for divisibility by 3, but here the digits can always eventually be added up to make 9.

Is 64152 divisible by 9?

$6 + 4 + 1 + 5 + 2 = 18$

$1 + 8 = 9$ so 64152 is a multiple of 9.

Divisibility by 11

This is a bit more complicated. Begin by separating out the alternate digits of the original number (starting from the left-hand side), and add them together.

Is 95428399 divisible by 11?

9 5 4 2 8 3 9 9

$9 + 4 + 8 + 9 = 30$

$5 + 2 + 3 + 9 = 19$

Now subtract the smaller of these results from the bigger one. If the result is 0 or a multiple of 11, the original number is divisible by 11.

$$30 - 19 = 11$$

so 95428399 is divisible by 11.

You could just use a calculator, of course.

π (3.14159265358979323 ...)

π (pi) is one of the better-known numbers in this collection, its importance having been drummed into us at school. It is, of course, the number that defines the relationship between the circumference and the diameter of a circle ($C = \pi D$, see illustration below), and it is named after the first letter of the Greek words for periphery or perimeter.

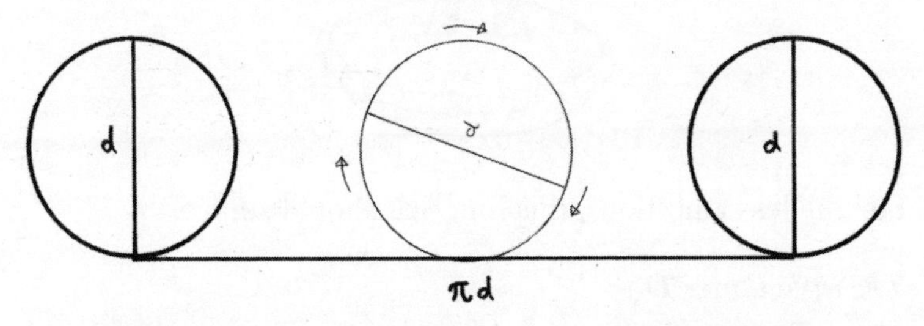

Strangely, though, π also appears in a vast number of other fields, particularly quantum physics and cosmology, which appear to have no relation to circles. It's even been the subject of legislation (see Indiana Pi Bill, p.49).

For all its scientific importance, π is perhaps more memorable for being a non-terminating, non-repeating decimal (an irrational (see p.121) and also a transcendental number). Technological advances in the late twentieth century allowed vast numbers of the digits of π to be calculated by computer, and the current record is about 1.2 trillion decimal places.

Despite the lack of any real point in doing so, memorizing the digits of π has become an almost obsessive pastime among some people. In October 2006, retired engineer Akira Haraguchi memorized 100,000 digits, beating his own previous record of 83,431 places. Pi devotees often use complex mnemonic devices in their efforts, such as *The Cadaeic Cadenza*, a short story by Mike Keith, the words of which correspond in length to the first 3834 digits of π.

For the less ambitious, this limerick should suffice:

The ratio C over D
When they're parts of a circle, is three
Point one four one five
Nine two six five three five
Eight nine seven nine three two three!

These π devotees often also celebrate Pi Day on 14 March (in America read 3/14), by eating π pie. Depressingly, this is one of the funnier running gags of the mathematical world.

THE EVOLUTION OF ZERO

To most of us, the idea of life without zero is inconceivable. From the most basic financial transactions to the circuitry of all electronic equipment (see binary system, p.150), the zero underpins every aspect of modern life. The value and meaning of zero are among the first things children learn.

But for many centuries, the idea of a number that represented nothing was counterintuitive and controversial, raising philosophical as well as mathematical questions. Europe's first mathematics, after all, was concerned mainly with counting and measuring solid objects, and you can't count to nothing. The story of zero is also largely the story of how the West adopted a positional decimal number system over the more primitive Roman one.

Around 3000 BC – first evidence of a primordial decimal number system in the Indus valley, in what is now India. The civilization of this area was enormously mathematically advanced, and remained ahead of others for centuries in its understanding of number theory.

mid-1000s BC – the Babylonians developed a rather complex sexagesimal (base-60) positional notation system, which required some placeholder to represent an unoccupied space. It could be used to distinguish, say, 101 from 11 but could not be called analogous to modern 0, because it was never written at the end of a number and was not considered to have a value in itself (and, consisting only of an empty space, could not be written on its own). Between 700 and 300 BC, the Babylonians developed symbols similar to the current one to represent zero, but still did not have a number resembling modern zero.

300 BC onwards – Hindu cultures adopted something closer to the modern zero and gave a Sanskrit name to the null value – *sunya* (the etymological root of the word zero).

The Greeks – by contrast, had philosophical difficulty with the idea of naming zero – it seemed contradictory that 'nothing' could be 'something', especially as the Greek mathematical tradition, unlike the Hindu one, was concerned largely with counting and measuring physical objects. Geometry was the real forte of the Greek mathematicians. This philosophical doubt eventually extended to medieval religious arguments about the existence or otherwise of a vacuum.

However, Greek mathematician Ptolemy, influenced by the Babylonians, developed the 'Hellenistic zero' in AD 130. He gave it a symbol similar to the current one – a circle with a long overbar. It would take a long time to spread to the rest of Europe, though, and Ptolemy, like the Babylonians, did not view 0 as a number in its own right.

The Romans – did not use a numeric zero symbol, preferring the word *nullus* or *nihil* (nothing), alongside their other numerals. The word was occasionally used alongside other numbers in tables (the oldest surviving example dates from *c.* AD 525), but the Roman system, which was used in the West for centuries, was not positional and had little need for a zero.

AD 498 – the now-ubiquitous decimal number system was by this time fully established in modern-day India. The mathematician Aryabhata described a decimal positional system where *'Sthanam sthanam dasa gunam'* – 'place to place, ten times in value' – but viewed the zero as an indicator of an empty position rather than a number itself.

AD 628 – Hindu mathematician Brahmagupta laid down a set of rules for zero, negative numbers and some simple algebraic rules, in his book *Brahmasputha-siddhanta.* These mostly correspond to those used today, but modern mathematics disagrees with his assertion that $0 \div 0 = 0$. Dividing by zero remained a thorny issue for some time (see p.136).

Around 665 – The use of zero as a placeholder was also popular in the Americas, in the vigesimal (base-20) system of the Long Count Calendar, which the Maya, among other civilizations, used to predict major celestial events so they could plan their sacrifices. It used a shell to represent 0, and was probably developed by the Olmecs and before, spreading to the Maya. However, this development appears not to have spread beyond Central America.

AD 820 – Brahmagupta's writings were introduced to the wider world, centuries later, by Al-Khwarizmi (around 790–840), a Persian mathematician who invented algebra (in his book *Al-Jabr wa-al-Muqabilah*) and gave his name to the algorithm. In doing so, he brought together Greek and Hindu knowledge of mathematics. Arabic numerals had developed over the previous two centuries from Hindu ones, and would eventually become the standard across Europe.

AD 976 – the first appearance of Hindu-Arabic numerals in Europe in the *Codex Vigilanus*, a historical record of Spain. Zero, however, was not included by the monks who compiled the book.

11th century – the Indian decimal numeral system reached widespread use in Europe through the Iberian Peninsula, brought by the Moors along with Hindu-Arabic numerals. The Italian mathematician Fibonacci (see Fibonacci sequence, p.140) had travelled to numerous trading ports and favoured the Arabic system, becoming instrumental in promoting it in his *Liber Abaci* of 1202. The new system had many conservative opponents in the Catholic Church – the word 'cipher' is derived from the Arabic *sifr* (zero or empty), possibly because the idea of the zero was so hard for ordinary Europeans to grasp that the term came to be used for anything unclear, like a coded message. Even then, zero was viewed less as a number in its own right than as a modifier digit that could distinguish 3 from 300. Arabic numerals, including zero, didn't become truly widespread until the development of the printing press.

1440 – Gutenberg developed the printing press, and with it came a new ability to disseminate information. The socioeconomic changes of the Renaissance led to enormous growth in trade, and a growing mercantile class used the new numbers. Use of Arabic numerals was no longer limited to mathematicians, but it took centuries for the zero to be fully accepted.

To present day – growing European control of most of the world, particularly economically, meant Hindu–Arabic numerals were spread worldwide.

Dividing by Zero

One of the most perplexing problems with zero, even now, is how to divide a number by zero. Taking a/0 to equal 0, as Brahmagupta did in 628, seems to suggest that 0 times itself could equal anything other than 0. Taking it to equal ∞, as Bhaskara did 500 years later, makes a little more sense, but there is the problem that infinite nothing is still nothing.

As always, computers cause as many problems as they solve – a computer required to divide by zero may generate an incorrect result, or (worse) expend all its resources trying to solve the impossible problem, and not have the resources for anything else. Many programs are designed to spot a divide-by-zero error before it can cause problems, but this was small consolation to the crew of the USS *Yorktown*, a US Navy cruiser whose computer network hit a divide-by-zero error. The network crashed, and with it went control of the ship's propulsion, leaving it paralysed in port.

A convenient, if perhaps rather unfair, solution, is simply to remove this calculation from the field of mathematics altogether. Recent computer mathematics standards require a division by zero to produce the result NaN – Not a Number – much like zero itself was all those centuries ago...

ROMAN NUMERALS

You probably at least half-know these already, but this does seem an appropriate place for a refresher on the system of ancient Roman numerals, although these days you're unlikely to have any pressing need to use or read them. Their modern use is largely confined to clock faces, statues and formal documents.

The decimal number system now in use worldwide uses positional notation, with the position of each digit indicating its significance (see Base, p.117, for a more detailed explanation). This system derived from a need for a number system that allowed complex theoretical mathematics to be performed.

Despite their prominence in engineering, architecture and trade, the Romans, unlike the Greeks, produced little in the way of abstract mathematical theory. The Roman numeral system is based essentially on counting things, and although Roman numbers are also letters of the alphabet, this was not their original source. The system appears to have begun with Etruscan shepherds, who carved notches into tally sticks when counting their flocks. To make the process easier, they would carve a second stroke with each fifth notch, two crossed strokes for each tenth one, and so on. Gradually, the Romans adapted the shepherds' system into one that used the letters of their alphabet.

I – one
V – five
X – ten
L – 50
C – 100
D – 500
M – 1000
Putting a little bar over the digit multiplies its value by 1000, and was common for numbers above 4000 or so:

$\overline{\text{IV}}$ – 4000
$\overline{\text{M}}$ – 1,000,000
$\overline{\text{MDCC}}$VII – 1,700,007

Unlike in our decimal system, the placement of the symbol has little to do with its significance (i.e. the M in MXI is not ten times bigger than the M in MI). The numerals, in most cases, simply add together, and are usually written largest first e.g. MLXVI. However, this is not always the case, and working out Roman numerals isn't always just a question of adding everything together. To make the writing process less cumbersome, the Romans would often express numbers that fell just short of one of the convenient quantities above as a larger number less a smaller number – because of their tally-based system, it was quicker to write 'fifty less one' than 'forty-nine'. This is done simply by writing the smaller number to the left of the larger one:

IV = 4 (though IIII was also common)
IX = 9
LD = 450
MCMLXVIII = 1968

There was no real method of writing zero. Any Roman needing to express this concept would have to write out the word *nullus* (meaning nothing). After the decline of the Roman Empire, the numbers continued to be commonly used until the introduction of modern Arabic numerals around the fourteenth century (see The Evolution of Zero, p.130).

THE FIBONACCI SEQUENCE

The Fibonacci sequence is a sequence of numbers, defined by the fact that each term is the sum of the previous two terms. Therefore the first fifteen terms are:

0, 1, 1, 2, 3, 5, 8, 13, 21, 34, 55, 89, 144, 233, 377

The sequence is named after Leonardo of Pisa, also known as Fibonacci (c.1170–1240), an Italian mathematician who was a major figure in spreading Arabic numerals to the rest of the world (see The Evolution of Zero, p.130) in his book *Liber Abaci* (the Book of Calculation, published in 1202). *Liber Abaci* also introduced the Fibonacci sequence as a solution to the following problem:

> A pair of rabbits is kept in an enclosure, supplied with food and water. After a month, this pair mates and produces a new pair of rabbits. Each new pair of rabbits begins producing offspring after one month, and produces exactly two every month after that. How many pairs of rabbits can be produced in each month over the course of a year, assuming each pair dies after producing two pairs of offspring?

The result is the Fibonacci sequence.

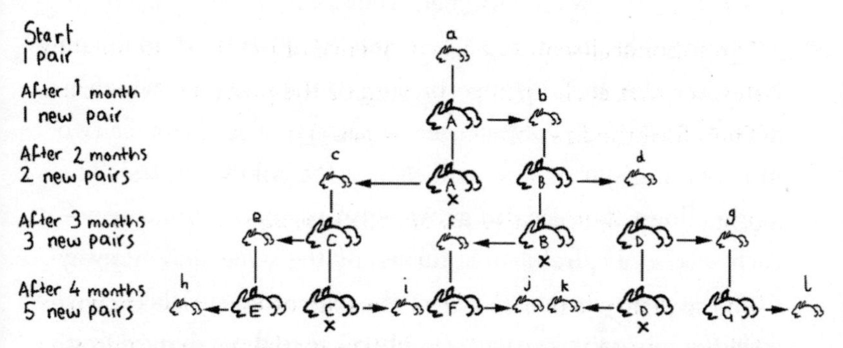

The Fibonacci sequence has attracted a great deal of interest because it crops up frequently in nature – even outside monogamous rabbits, which actually occur rather rarely. The Fibonacci sequence also appears in sunflowers, in the spiral pattern of the seeds in the centre (which usually grow in formations of 55 clockwise and 89 anticlockwise spirals of seeds – both Fibonacci numbers). A similar pattern exists in other flowers, pine cones and pineapple fruitlets. It gives the most efficient, even distribution of seeds in the space available.

The sequence is also strongly linked to the Golden Number (see p.44) – the aforementioned seeds emerge from each other at an angle of about 222.5°, or 360° divided by the Golden Number (the angle is also often given as 137.5°, which is 360° - 222.5° – this is because the pattern of seeds forms two intertwining spirals, one with each of the golden angles). The angle allows the seeds to grow outwards in a regular way – each seed in a spiral arm is turned by the same amount away from the previous one. Patterns of squares and spirals derived from the Fibonacci sequence closely relate to those derived from the Golden Number. As with the Golden Spiral, sightings of this phenomenon in nature can be rather spurious.

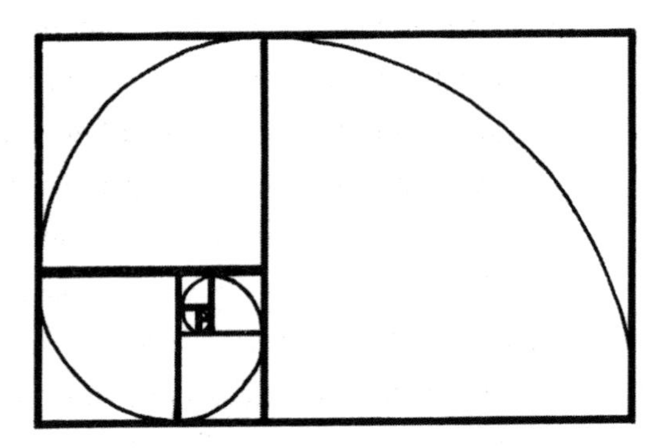

I (√-1) THE IMAGINARY UNIT

Finding the square root of a positive number, even one that humans would find difficult and lengthy, is now an extremely simple process for even the most basic calculators. The square root of a *negative* number, however, remains entirely impossible to find. Any positive number squared will produce a positive result, and so will the square of any negative number (in maths, two negative numbers multiplied together will produce a positive result: $-3 \times -2 = +6$).

But i – the square root of -1 – is not only mathematically existent but rather useful too. For i is the basic unit of the imaginary numbers, an entire set of numbers centred around negative square roots, which can be manipulated according to many of the same rules as real numbers, and are often combined with real numbers to form complex numbers (3 + 2i, for instance, is a complex number. π (see p.129), however, is just a complicated one).

The seeming impossibility of the imaginary numbers meant that for years very few mathematicians would approach them. Even ordinary negative numbers seemed rather outlandish. Many figures of Renaissance mathematics dismissed the idea of a negative square root as absurd, but one, Girolamo Cardano (1501–76), acknowledged their existence and usefulness in solving cubic equations in his book *Ars Magna* (the Great Art). As well as being a mathematician, Cardano was a physicist, an inventor of many devices, including the combination lock, and an incorrigible (and not wholly honest) gambler – and hence, presumably, no stranger to negative numbers. Eventually, though, he was imprisoned in 1570 for heresy (he had published Jesus' horoscope) which left him looking rather an ars magna.

After his death, Cardano's work was continued by his colleagues, but imaginary numbers were not fully accepted for many years. After all, one might well wonder what relevance imaginary numbers, being imaginary, could have to the real world. Well, you've probably seen sound waves represented like this:

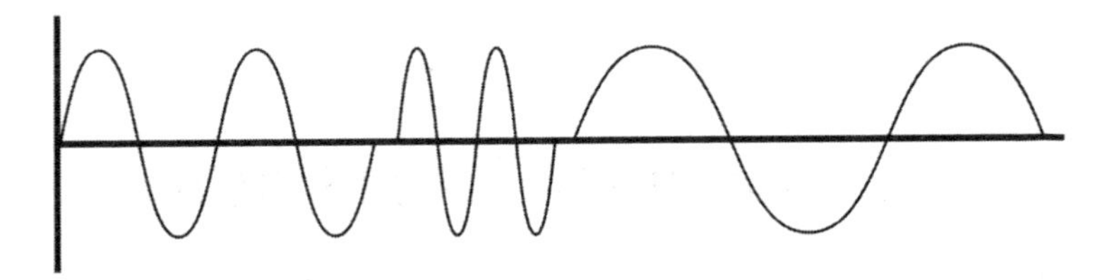

This is known as a sine wave, because it's based on the sine – a trigonometric function you may well remember from school. In a right-angled triangle, the sine of any angle is equal to the length of the side opposite the angle divided by the hypotenuse (the side opposite the right angle is the hypotenuse, thus sin 90 = 1).

All sounds are made of one or more sine waves: a louder sound will have higher and deeper peaks and troughs, while sounds at higher frequencies will appear more horizontally 'squashed' than lower ones. However, sound is by no means the only appearance of sine waves – they also appear in other oscillations like the voltage of alternating current systems, in radio waves and light waves. Even the literal waves of the ocean are basically sine waves, though I wouldn't bother bringing graph paper to the beach with you.

Without going too deeply into the complicated maths of this bit,* the relationship between i (which, confusingly, is often written j in engineering contexts) and π allows

* For a more mathematically rigorous but still very accessible explanation, I'd recommend the final chapter of David Acheson's *1089 and All That.*

sine waves to be expressed in relatively simple ways using complex numbers, reducing massively the work involved in calculating the maths of oscillations. Hence, a number that seems mathematically impossible is still enormously helpful to engineers and scientists everywhere.

MISLEADING STATISTICS

Numbers exert enormous control over our lives as statistics, where they are cited to support assertions and determine policy. All too often, though, the statistics are misleading, whether unintentionally or otherwise.

Any credible study should have a large enough sample size – that is, enough things or people being studied – to rule out anomalies. The smaller the sample, the higher the chance of results that don't reflect the population at large. Furthermore, a survey that overrepresents a particular group of people, whether by how it's run or by pure chance, might show a bias. For instance, a political survey conducted in expensive suburbs in the south of England will probably be biased towards the political right, while the same study conducted in the middle of a northern city would get the opposite result.

It's all in the selection

Selection bias is also an issue, occurring when the sample selected is inherently skewed. There are many ways of doing this – one example might be surveying gym customers about their views on health, which would probably be unrepresentative as gym users tend to be more health-conscious. The time period under scrutiny is also important: a short-term survey is not very useful for discussing long-term trends.

Self-selection bias is especially important in the media – surveys, especially political surveys – can be biased by the group of people who actually bother to do them. Website surveys are particularly problematic, because most websites have a specific audience and the cheapness of running them makes them useful to the press. Combine that with the political agendas of most newspapers, and you have an easy system for a newspaper to statistically support any claim it cares to make.

In 2008, a right-wing British tabloid newspaper ran a poll about reinstating capital punishment. Given the politics of its readership, the poll would probably have been skewed anyway. However, those who bothered to vote on the website and in telephone polls were even more likely to be in favour – as the process of voting requires action from the voter (rather than passively having someone telephone them). Hence the poll is more likely to be accessed by those who want a change. Newspapers, particularly tabloids, often use emotional, one-sided coverage to sway those voting in these polls, just as they do in general elections. The capital punishment poll eventually found that 99 per cent of readers favoured its reinstatement, while more objective surveys suggest that only around 45 per cent of Britons would agree.

It's the way you ask them

Loaded questions are an excellent way to get the data you want. In the US, this has reached an incredible height with the 'push poll' – an outwardly objective survey that hides unfounded accusations in its questions.

A classic example took place in the 2000 Republican presidential primary in South Carolina, where voters were asked by phone 'Would you be more likely or less likely to vote for John McCain for president if you knew he had fathered an illegitimate black child?' As well as appealing to racism in one of the ugliest ways possible, the poll question also uses sneaky verbal loopholes – at no point does the caller accuse McCain directly, but the wording of the hypothetical question suggests the information is correct. (The accusation was lent weight by the presence on the campaign of McCain's daughter, Bridget, who was adopted from Bangladesh.)

Averages (see p.116) can also be misleading. The mean, in particular, can be easily influenced by unusually high and low values. In a group of five people, where four earn £25,000 a year and one earns £250,000, the mean salary is £70,000, which is evidently unrepresentative. For this reason, median salaries are generally used for official purposes, though these data are far from perfect. Almost all official data, for instance, excludes people who are working illegally.

Lies, damned lies, and statistics

But all this is nothing compared to the way statistics are used to draw conclusions. One has to remember, when looking at any set of statistics, that correlation does not imply causality – just because two factors coincide doesn't mean one causes the other. For instance, we could consider that generally, people who wear glasses tend to be considered more intelligent. Does that mean glasses make you more intelligent or that intelligence somehow affects the eyes? Probably not. It seems far more likely that either more intelligent children are more likely to spend time reading at a young age, which may affect their eyes, or (as has recently been suggested) intelligence in small children is linked to visual exploration of the area closest to them.

This is a mistake even the most eminent scientists can make. In the study of malaria, for example, (which gets its name from the Italian words *mala aria* – 'bad air'), for centuries European scientists mostly assumed that, because cases of the disease occurred mainly around swamps and marshes, the smelly air to be found in these places was what caused it. Although moderately effective treatments were developed, millions of people died because of an assumption that held back progress. It wasn't until the end of the nineteenth century that it was conclusively proved that mosquitoes transmitted malaria. This had been suggested long before – even the ancient Egyptians seem to have known it. Worldwide mosquito control efforts over the twentieth century – including mass pesticide use and supplying mosquito nets – have massively reduced the incidence of malaria, although it remains a deadly disease in Africa.

This sort of misinterpretation isn't confined to science, or to numerology. It seems to me to be a fundamental problem with people's mental software. We evolved to recognize patterns in data of all kinds – whether it's the hunter recognizing the familiar shape of a deer in a dense forest, or a mathematician spotting a similarity between two complicated sequences of numbers – but with that comes the danger of seeing patterns that aren't there. In numerology, this means coincidentally encountering a particular number a few times in a day and assuming it's somehow auspicious. In statistics, the danger in identifying a nonexistent pattern can be much more serious.

THE BINARY SYSTEM

The binary number system contains just two digits – 1 and 0 – but is at the core of all electronic devices. It is the simplicity of the system that allows it to be used by machines – an electronic pulse is high or low voltage, one or zero – but by combining many electronic 'bits', hugely complex data can be expressed.

Just as the decimal system works on powers (see p.118) of ten, the binary system works on ascending powers of two. Any positive integer can be expressed as a sum of ascending powers of two, with only one of each power at most being needed. Three, for instance, is written in binary as 11 – the sum of 2^1 and 2^0 – while 17,689 appears as:

100010100011001

Numbering the digit positions by powers of two:

1	0	0	0	1	0	1	0	0	0	1	1	0	0	1
14	13	12	11	10	9	8	7	6	5	4	3	2	1	0

2^{14} = 16,384 +

2^{10} = 1,024 +

2^{8} = 256 +

2^{4} = 16 +

2^{3} = 8 +

2^{0} = 1 +

= 17,689

Or, as the age-old nerdy joke goes: There are only 10 kinds of people in the world – those who understand binary, and those who don't.

In electronics, the binary impulses are usually interpreted by a system of logic gates – tiny components that receive one or two binary inputs, each either 1 or 0, and output a binary pulse according to the 'logic' involved. Millions of logic gates can be fitted onto a single computer chip, and this mind-bogglingly complex system of inputs and outputs is essentially what gives a computer its power. This was true even before electronics was invented – Charles Babbage's Analytical Engine, a mechanical calculating device designed from the 1830s onwards and often considered the forerunner of modern computers, would have relied on a system of purely mechanical logic gates, had it ever been fully built.

The byter bit

The binary system is perhaps most obvious in computer memory storage. Each binary bit (a one or zero) is part of a byte* – a group of eight bits, traditionally enough to represent a single character of text – of which 1024 form a kilobyte (KB), though this is sometimes also given as 1000†. A megabyte (MB) is formed by 1024 kilobytes, or 1048576 bytes, and 1024 megabytes make a gigabyte (GB). So, to put all this into context, this brief explanation of the binary system has made a 69-kilobyte file, which in turn is made up of about 561,152 binary bits; a 4-minute MP3 is made up of about 32 million bits, and an 80-gigabyte iPod hard disk can be reduced to about 687 billion ones and zeroes.

---

* The name 'byte' has given rise to a number of hilarious nerd puns for other small units of memory, such as a nybble (4 bits) or the dynner (32 bits). The Jargon File at http://catb.org/jargon/ is a good source for many, many more examples of this kind of computer lore.

† There's some confusion about exactly how a kilobyte should be defined. The Greek prefixes 'kilo', 'mega' and 'giga' properly mean thousand, million and billion respectively, and it's natural enough for people to see computer memory in these decimal terms. However, the physical binary storage is generally in multiples of 1024 bytes because that's a power of two. Although this may seem insignificant, it can cause fairly major differences as the storage units get bigger – a difference of 24 bytes (2.4 per cent) at the kilobyte level increases to around 73MB (7.3 per cent) at the gigabyte level. This difference can be exploited for sales purposes, as a hard drive with around 250,000,000,000 bytes of memory can be advertised as being 250GB, whereas in fact it's about 233GB. In 2000, in an effort to correct this problem, a new system of prefixes that specifically referred to the relevant powers of two was introduced, working from the kilo-binary-byte (KiB, always 1024 bytes) upwards.

ODDS AND ODDITIES

Probability is the branch of mathematics concerned with measuring the likelihood of a given outcome, which is usually expressed as a decimal, percentage or fraction. Where the outcome is random, like rolling a dice, the probabilities are evenly distributed, i.e. the probability of any given number is the same as any other (in this case, $\frac{1}{6}$). Where there are more possibilities for one outcome than another (for instance, you're more likely to pick an E from a Scrabble bag than an X), the probability will be higher for that outcome.

Probabilities of multiple events in sequence can be worked out by multiplying the probabilities of the individual events – the probability of rolling a 6 on a dice three times in succession is $1/6 \times 1/6 \times 1/6 = 1/216$ or about 0.5%. The probabilities resulting from each individual event or sequence of events always add up to 1 in the end.

The easiest probabilities to work out are those for random things where the chance of each outcome is equal – for instance, the probability of a tossed coin landing on heads is 0.5, which is also the probability of it landing on tails. There is a tiny probability of it landing on the edge and staying on there – but we'll ignore that for the moment to look at a couple of interesting applications which show how counterintuitive probability can be.

The birthday problem

If there are 23 people in a room, what is the probability of two or more of them sharing the same birthday? One might think the probability is very low – after all, there are 365 days in the year, and only 23 people. In fact, though, it's more likely than not. Why is this?

First off, we'll assume for the sake of simplicity that all the birthdays are equally distributed throughout the year, and that no one was born on a 29 February. The probability of a person being born on a given day, therefore, is $\frac{1}{365}$, since there are 365 days in the year and a person is equally likely to be born on each one of them.

The trick here is to remember that it doesn't *matter* which people, or how many people, share a birthday, or which day it is. We could try and work out the probabilities for every possible grouping and each day, but that would take forever, and there's an easier way. We know that the probability of two or more people sharing the same birthday (which we'll call P(s)) and the probability of no two people sharing the same birthday (P(ns)) must add up to 1, because the probabilities of all the outcomes of a given event always add up to 1.

Therefore, we can simply calculate the odds of no two people sharing the same birthday, i.e., the probability of every single person being born on a different day, and then subtract that from 1 to get the probability that at least two people will.

Imagine now that you're going through the room carrying a clipboard and wearing a white coat, asking each person their birth date, and assuring them they'll be released once the study is complete. The first person in our big probability calculation could be born on any day, so the first term is $^{365}/_{365}$. The next person can't have the same birthday as the first, but still has 364 possible birthdays, so the second term is $^{364}/_{365}$. And so it goes on for all 23 of them, as each person has to have a birthday from a slowly narrowing range of them.

$$P(ns) = {}^{365}/_{365} \times {}^{364}/_{365} \times {}^{363}/_{365} \times \ldots \times {}^{343}/_{365} = 0.493$$

$1 - 0.493 = 0.507$, or 50.7%.

Hence, it's more likely than not (just) that two or more people will share the same birthday. As the number of people increases, it quickly gets more likely, rising to a 99 per cent chance for a group of 57 people. However, it's only absolutely certain when there are 366 people in the room, as it's still just possible for every person to have a different birthday until there are more people than days.

The Monty Hall problem

This next one's a bit trickier, and has been around in some form for quite some time. It's named after Monty Hall, former host of the American game show *Let's Make a Deal.* The show provides the template for this problem, which has apparently fooled even eminent scientists.

The rules of the game show are as follows. There are three doors. Behind one is a brand-new sports car, but behind the other two are dud prizes (traditionally goats, for some reason). The car is randomly placed. In the first round, you, the player, must choose one of the doors, with no knowledge of what's behind any of the doors.

Then begins the second round. Our host, who knows which door is the winning one, makes things a bit more complicated by opening one of the other two doors at random – never the winning one or your chosen door – even if you picked the winning door in the first round, that won't be revealed until the end. You are now allowed to stick with your original door or change your choice to the other unopened door. The door you choose now will be opened to reveal your prize.

Given that you're now presented with two doors, one of

which conceals a car and the other a goat, it might seem that your chances of winning are equally likely, and it doesn't matter what you do. In fact, you should always switch.

The trick here is to remember that the first two rounds are connected, rather than being independent events, and that you don't know which door's correct until both rounds are over. With that in mind, we'll consider the probabilities.

You have a ⅓ chance of choosing the winning door in the first round. If you do so, switching doors in the next round would cause you to lose, no matter which of the other two doors the host opens. However, you have a ⅔ chance of picking a losing door in the first round. In that case, the host will have to open the other losing door. Now, switching doors will guarantee you the car.

You have no idea, of course, whether the first door you've chosen is correct, but there's a ⅔ chance it wasn't, and this probability is not altered by the door the host opens. It's twice as likely as not that you chose the wrong door first time, and hence twice as likely that switching will win you the car. It's no guarantee of victory, but switching doors in the second round should win ⅔ of the time.

JOHN NASH AND GAME THEORY

The Cold War is chiefly remembered for the stockpiling by both sides of terrifying quantities of nuclear weapons. These weapons were never used against their targets, except as bargaining chips in a sinister and Byzantine doctrine known as 'Mutually Assured Destruction'. American strategists devised complex predictions of possible Russian first strikes, and planned retaliatory strikes to ensure Western victory (insofar as anyone wins in a global nuclear war, anyway), no matter which strategy the USSR pursued. These plans relied on a view of humans as coldly rational, scheming, self-interested beings with little interest in altruism or a greater good.

At the heart of this worldview was the American mathematician John Forbes Nash (b. 1928) who produced pioneering work in the field of game theory – a branch of mathematics dedicated to analysing human behaviour mathematically, originating in attempts to predict the outcomes of chess and card games. Nash, along with a number of colleagues, advanced game theory and began applying it to global politics while working at the RAND Corporation, a military think tank, during the 1950s. For Nash in particular, all human interactions could be viewed in terms of Machiavellian self-interest.

The game theorists developed various game scenarios designed to mimic the behaviour of these self-interested hypothetical people – a well-known example is the Prisoner's Dilemma (see p.160), invented by Nash's colleagues at RAND. They claimed that such behaviour could be applied not simply to games, but to every aspect of human existence, including the Cold War. By viewing the USSR as implacably hostile but also self-interested, they led RAND to promote the doctrine of mutually assured destruction. Like the Prisoner's Dilemma, each side had the opportunity to betray and annihilate the other, but there would be serious consequences in doing so.

But despite their influence on the policy of the Cold War, the worldview of game theory seems to have major flaws. Tests of a game called So Long Sucker (as the name implies, the game is won by forming and then breaking alliances) on secretaries at RAND were failures – loyalty ended up trumping their desire to win the game.* Nash, whose life inspired the 2002 film *A Beautiful Mind*, was diagnosed with paranoid schizophrenia in 1959, but his work remained influential. Game theory is not simply concerned with nuclear paranoia

* Interestingly, it was the same compassion and initiative that saved the world when it was closest to destruction. On 26 September 1983, a Russian army engineer named Stanislav Petrov was on duty at a surveillance station when an alarm went off, as though America had launched a strike. Petrov realized it could be a false alarm (as it turned out, a computer error) and that the future of the world rested with him. He disobeyed protocol by refusing to alert the Soviet command, preventing nuclear apocalypse but destroying his own career.

– it has important applications in economics, political science and computer science. Nash, now fully recovered, received a Nobel Prize in Economics for his work in 1994, and some people have claimed that his work in game theory has created a world ruled by numbers.

The Prisoner's Dilemma

Two bank robbers, A and B, have been arrested by the police. The police have found illegal firearms in both their houses, and it seems clear that these are the weapons recently used in a major bank hold-up. Despite the discovery of the guns, however, the police have failed to find sufficient evidence to secure a conviction for the robbery. They intend, by separating the two prisoners, to force one to betray and testify against the other.

Each prisoner is thereby presented with a dilemma. If A betrays B, he will go free and B, as the only one to be convicted, will get a long ten-year sentence. A, however, realizes B is probably thinking the exact same thing about him. If neither betrays the other, they will both get a few months inside for possessing the illegal guns. If they both betray each other they'll both be convicted for the robbery, and end up with five-year sentences – not as long as if only one were betrayed, but far worse than if neither betrayed the other.

The maxims of game theory dictate that the best thing for each prisoner, in terms of personal gain, is to betray the other. At best, you get away with the crime. Given that the other prisoner is probably also intending to betray you, you gain a reduction in your sentence by betraying him.

BILLIONS

There is great confusion surrounding the usage of the word 'billion'. The problem is that the United States uses the term to mean 1,000,000,000 – one thousand million, or 10^9, while most of Europe uses it to mean a million million – 1,000,000,000,000 or 10^{12}. The European billion seems to make more sense given the 'bi' prefix, which would suggest a billion is a million to the power of two, a trillion is a million to the power of three, etc. In most of these countries, the word 'milliard' is used for a thousand million and sometimes 'billiard' for a thousand million million, 'trilliard' for a thousand million million million, and so on.

The difference between them stems from two different systems of working with very large numbers – the Long Scale used by mainland Europe, and the Short Scale used by America and most English-speaking countries. The Long

Scale assigns a different word to numbers over a million that are a million times larger than the previous one, while the Short Scale does so using a factor of a thousand – a Long Scale trillion is 10^{18}, while a Short Scale one is only 10^{12}. As the numbers get bigger, so does the difference – by a factor of a thousand each time.

The UK used what we now call the Long Scale from around the sixteenth century, and its adoption by Germany spread it around to most of Europe. France, where the original Long-Scale billion and trillion were invented, then switched to the Short Scale in the early eighteenth century. This usage spread to French colonies in America, although France officially switched *back* to the Long Scale in 1961. The UK stayed with the Long Scale officially until 1974, when the Wilson government announced a conversion to Short Scale numbers in official literature. However, even today people in Britain are likely to confuse the two.

THE WHEAT AND CHESSBOARD PROBLEM

This problem, often used to demonstrate the speed of exponential growth, has its origins in a story of India in about the sixth century AD. Supposedly, an Indian mathematician called Sessa was the inventor of the game we now know as chess (an early version known as *Chaturanga*) and was called before his king to demonstrate his invention. The king was so pleased, he offered to give Sessa any reward he could think of.

Sessa asked the king only for wheat (rice in some versions) – one grain on the first square of the chessboard, two on the second, four on the third, eight on the fourth and so on, doubling each time until he reached the last of the 64 squares. At first the king derided this seemingly paltry request, but as the numbers were added up he realized it was enough wheat to bankrupt him completely – 18,446,744,073,709,551,615 ($2^{64}-1$) grains. Even with modern methods, producing that much wheat would require harvesting all the arable land on Earth eighty times over.

Sessa can't have received anything like his entire reward, and one variant of the story has the king getting an exacting revenge – out of feigned concern for Sessa, the king asked him to count every grain he received, to make sure he wasn't being short-changed. Although the Sessa story is probably the most common, there is also a version of the legend set in the Roman Empire. None of it – except the numbers – should be taken as true.

THINKING FOURTH-DIMENSIONALLY

As physics made strange and exciting new strides in the twentieth century, it became clear that there were problems with traditional scientific ideas, which viewed the universe as existing in three dimensions of space. Isaac Newton (1643–1727), whose theories formed the basis of physics until the twentieth century, saw time as flowing at a constant pace, regardless of anything else in the physical universe. For everyday life, where we rarely if ever travel at significant fractions of the speed of light, this works very nicely, and for a long time the idea was not challenged.

However, the experience of time was shown by Einstein's theories of relativity not to be independent of movement in the traditional physical world: time appears to pass more slowly the faster the observer is travelling, and this process would be especially pronounced for someone moving close to the speed of light. The experience of space and time is relative, depending on the viewer's frame of reference. A well-known thought experiment involves a man travelling at nearly the speed of light for what seems to him to be a few minutes, only to return to Earth to find that decades have passed. Although travel at those speeds is obviously impossible at present (as well as presumably leaving the subject jetlagged to a degree far in advance of irregular sleep, and in fact well into the realms of existential horror), the effect has been observed with atomic clocks on board space shuttles. Even someone walking would experience time slightly differently to someone standing still, although the difference would be immeasurably tiny.

Anyway, all this suggests that time has to be related to the original three spatial dimensions in a much more complex way. With Einstein's Theory of Special Relativity as his starting point, the physicist Hermann Minkowski came up with a 4-dimensional model of the universe in 1908, with time as the fourth dimension. Space and time can then be viewed as a 'continuum' and, crucially, time can be distorted like space. Gravity can cause curvature in space-time, particularly where singularities (see Infinity, p.168) are involved. Light is known to bend around planets and stars because their gravity distorts space-time.

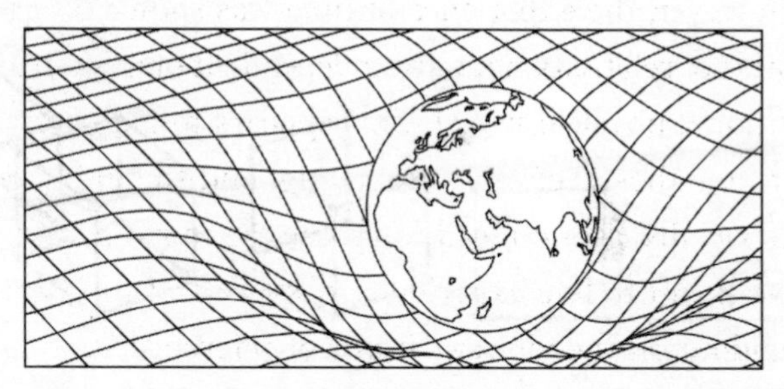

Another strand of mathematical theory, beginning some decades before, is concerned with a geometric fourth dimension of space, which is unrelated to time. The fourth dimension is perpendicular to the existing three, but as three-dimensional beings with basically two-dimensional vision we couldn't actually see it. The concept is difficult to visualize (although some physicists and mathematicians have claimed to think in four dimensions), but we could view it in terms of regular shapes. If we consider a straight line with one dimension (length) to be representative of a one-dimensional universe, a square representative of a two-dimensional one (since a square has two dimensions – length and width), and a cube (which possesses length, width and height) to represent a three-dimensional universe, the four-dimensional model is represented by a tesseract, or 4-dimensional hypercube. The tesseract is difficult to visualize and to draw, but could be considered like this: if a 3-dimensional cube is made of two 2-dimensional squares linked in a third dimension (with a total of 6 square faces), a 4-dimensional tesseract is made of two 3-dimensional cubes linked in a fourth dimension, creating a total of eight cubes.

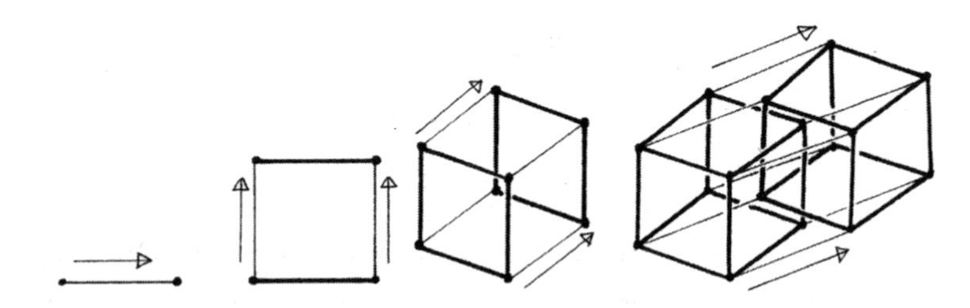

The 4-dimensional tesseract shown above does not look to us like a regular solid. But then, neither in fact does the cube look exactly like a regular cube – the edges that lead away from the viewpoint are slanted at a mathematically incorrect angle to conform to perspective, which is the method we have used to depict three-dimensional objects on a two-dimensional plane, ever since perspective painting was popularized during the Renaissance.

Similarly, we have to bend a few rules to display a 3D object as a 2D image, and the same goes for a 4D object. If we could see the tesseract in four dimensions, all eight cubes would look perfectly regular. As it is, they look slanted as below:

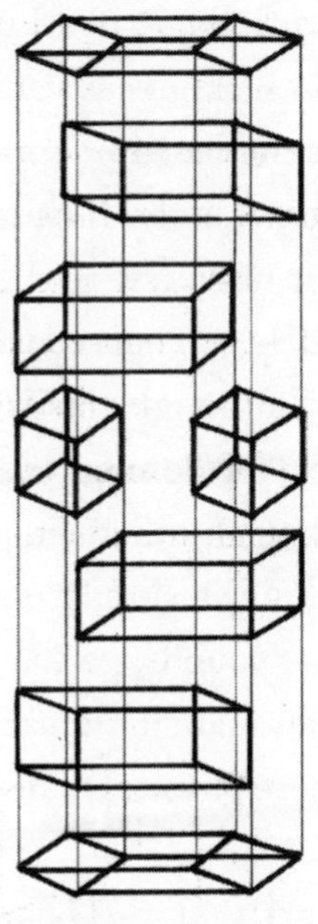

What use is a fourth spatial dimension anyway? One major application is in graphing functions that involve complex numbers (see p. 143), but that sort of thing is really a very long way outside the scope of this book. From our perspective, it just seems rather an interesting idea – a four-dimensional being would in theory be able to see everything in 3D space, and would be able to see behind and inside objects obscured to us.

∞ (INFINITY)

Infinity is perhaps not a number in a true sense, but the mathematical representation of all that is inexhaustible and everlasting. Essentially, it is a number that can never be exceeded (children's claims of there being an 'infinity plus 1' have not been subject to serious analysis – infinity plus anything is infinity), and which occupies a bizarre and fascinating place in our universe.

The universe is widely considered to be infinite in size and constantly expanding, although this is by no means certain. It may have some sort of defined edge, or the edges may somehow join up with each other – a person traversing one edge of the universe might simply end up on the opposite edge. As bizarre as that sounds, perhaps even stranger is the concept of the gravitational singularity – a point of infinite mass and zero size, where physical matter is compressed down almost to nothing.

There is a singularity at the heart of each black hole (thought to be the result of large stars collapsing in on themselves) whose gravity is so strong not even light can escape it, which is why such a phenomenon would appear black. Anyone entering a black hole would be subjected to incredible gravitational stresses, described by Stephen Hawking and others as like 'being turned into spaghetti', or sometimes referred to as 'noodlizing'. What happens after that is unknown, but unlikely to leave anyone alive. It is thought, however, that there is an enormous black hole at the heart of our galaxy and every other, and the gravitational pull of the black holes holds the galaxies together.

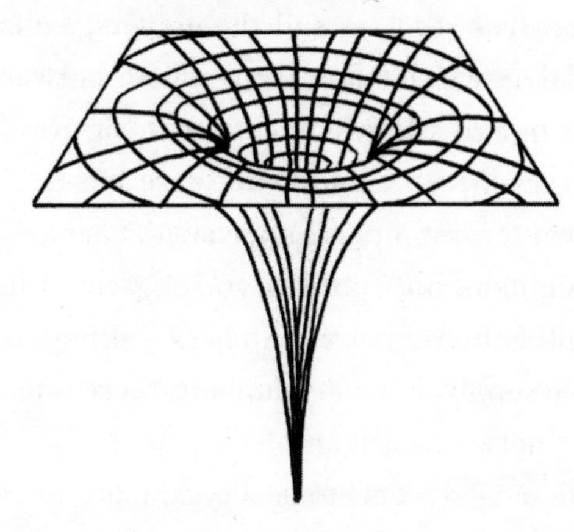

The entire universe is thought to have started off compressed into a singularity before the Big Bang. At this moment, around 14 billion years ago, the singularity exploded outwards. Superheated matter and energy shot outwards in all directions, and would gradually form everything there is in the universe today. It took hundreds of thousands of years even for the first atoms to form, and 100 million years for the first star to start shining. Our sun only formed after about 8 billion years. Even today, galaxies are still moving away from each other, suggesting the universe is still expanding.

In temporal terms, the universe is probably not infinite. Current theory suggests that in about another 100 trillion years, the universe will simply expand too far and cool down too much. Stars will burn away and galaxies fall apart in a 'Big Freeze', which will leave the universe dead, lacking the energy to support life. What will happen after that is another mystery, but the Big Freeze (also known as heat death) is not entirely certain. Other theories include a Big Crunch (the expansion process reverses and the universe collapses back into a singularity, which might then explode outwards to form another universe in a Big Bounce) and a Big Rip (continued expansion tears the fabric of the universe apart).

In numbers too, infinity is important. There are an infinite number of numbers, both positive and negative. Interestingly, there are different infinities of numbers – although there's an inexhaustible supply of whole numbers, there is an infinity of non-integers between each one.

This endless void seems to be a good place to end.

The Infinite Monkey Theorem

Back on Earth, infinity has found a place in the Infinite Monkey Theorem – the idea that, if you sit an infinite number of monkeys or other primates at typewriters, and leave them adequately fed and watered, one of them will eventually type out the complete works of Shakespeare, just by randomly pressing keys. The extreme improbability of any monkey doing that is counterbalanced by there being infinitely many monkeys (and, in some versions, infinite time), so one of them is bound to do it. The theorem was developed in 1913 by Émile Borel and Arthur Eddington, but wasn't tested until an experiment conducted in 2003 by researchers at Plymouth University. They aimed to investigate the Infinite Monkey Theorem by placing a computer with six crested macaques in a zoo enclosure. The results were, unsurprisingly, of little literary merit, with a stream of unconnected letters going on for several pages. The computer-illiterate primates also attempted to destroy the computer with a rock, as well as using it as a toilet. Some have claimed this invalidates the Infinite Monkey Theorem, but surely without infinite monkeys you can't ever really know – basic probability tells us that even extremely unlikely things will still occasionally happen. Some commentators, however, have suggested that the Internet proves that no matter how many monkeys are bashing away at keyboards, they still won't produce anything worthwhile.

FURTHER READING

I've made use of a number of other books and websites in researching this book. First, a couple of very wide-ranging books focusing heavily on the cultural significance of numbers – *The Book of Numbers* by Tim Glynne-Jones and *Numbers: Facts, Figures and Fiction* by Richard Phillips – proved to be very convenient references for an enormous range and number of facts, and I'd encourage anyone looking for a wider overview to seek them out. Also helpful were the amateur etymology websites Word Detective (www.word-detective.com) and World Wide Words (www.worldwidewords.org), while Snopes.com, as always, helped clarify a number of apocryphal stories, like the 'weight of the soul' (see p.37) or social security number 078-05-1120 (p.48). Michael Levin's book *The Complete Idiot's Guide to Jewish Spirituality and Mysticism* has an excellent chapter explaining how gematria works.

On the more mathematical side, *The Penguin Dictionary of Curious and Interesting Numbers* by David Wells is a superbly comprehensive guide to the mathematical properties of a great many different numbers, often providing excellent potted histories of the number in question. Similarly, the Wolfram MathWorld website (http://mathworld.wolfram.com) is a valuable resource for the lay person looking up mathematical terms and concepts.

David Acheson's book *1089 and All That* was extremely useful to me, and I'd recommend it even to the most maths-phobic as a wonderfully light and accessible introduction to some fascinating concepts of advanced mathematics. Finally, although their concern is less with specific numbers than with applied mathematics, Rob Eastaway and Jeremy Wyndham's books *Why do Buses Come in Threes?*, *How Long is a Piece of String?* and *How Many Socks Make a Pair?* should be of interest to anyone who enjoyed this book. (And even if you didn't, you bought it and it's too late now.)

OTHER TITLES AVAILABLE IN THIS SERIES:

i before e (except after c): old-school ways to remember stuff

by Judy Parkinson

978-1-84317-658-9

A Classical Education: the stuff you wish you'd been taught at school

by Caroline Taggart

978-1-78243-010-0

I Used to Know That: stuff you forgot from school

by Caroline Taggart

978-1-84317-655-8

My Grammar and I (or should that be 'Me'?): old-school ways to sharpen your english

by Caroline Taggart and J. A. Wines

978-1-84317-657-2